The Microbiome Miracle

How Tiny Microbes Shape our Inner Ecosystem

B.Sc. BEN MARFIELD

Table of Content

Book Introduction

The human body is a remarkable ecosystem, home to a vast and diverse community of microorganisms that play a crucial role in our health and well-being. These tiny microbes, collectively known as the microbiome, have co-evolved with us over millions of years, establishing an intricate and symbiotic relationship that profoundly influences our physiological processes, immune function, and even our mental and emotional states.

The world of microbes within us is a fascinating and complex one, with each individual harboring a unique microbial signature that is shaped by a multitude of factors, including diet, lifestyle, environmental exposures, and genetics. This invisible world, once overlooked and underappreciated, has now become the subject of intense scientific inquiry, as researchers unravel the profound implications of the microbiome on human health and disease.

At the heart of this exploration lies the intricate interplay between the trillions of microbes that colonize various sites within our bodies and the host cells they interact with. From the dense and diverse microbial communities that reside in our gut to the specialized microbial populations that inhabit other niches, such as the skin, vagina, and oral cavity, each of these ecosystems plays a vital role in maintaining our overall well-being.

The gut microbiome, in particular, has garnered significant attention due to its far-reaching effects on digestion, nutrient absorption, immune regulation, and even brain function. This intricate network of microbes, comprising bacteria, fungi, and viruses, is intricately linked to the health of the gastrointestinal tract and has been implicated in a wide range of conditions, from inflammatory bowel diseases to metabolic disorders and neurological conditions.

Beyond the gut, the vaginal microbiome plays a crucial role in reproductive health, influencing everything from fertility to the risk of infections and preterm birth. Similarly, the skin microbiome acts as a protective barrier, helping to maintain the delicate balance of our immune defenses and preventing the colonization of harmful pathogens.

The oral microbiome, often overlooked, is a gateway to whole-body health, with its composition and diversity linked to conditions ranging from dental caries and periodontal disease to systemic inflammation and cardiovascular health.

Moreover, the microbiome's influence extends beyond the individual, shaping the microbial landscapes of our homes, workplaces, and healthcare facilities. Understanding the dynamics of these environmental microbiomes can have profound implications for disease prevention and infection control strategies.

As we delve deeper into the intricacies of the microbiome, we are also uncovering the profound impact that our modern lifestyles and medical interventions have on these delicate ecosystems. Antibiotics, while invaluable in treating bacterial infections, can also disrupt the delicate balance of our microbiome, leading to dysbiosis – a state of microbial imbalance that has been linked to various chronic illnesses.

Similarly, our dietary choices and sedentary lifestyles can profoundly influence the composition and diversity of our microbial communities, potentially contributing to the rise of conditions like obesity, type 2 diabetes, and inflammatory disorders.

In response to these challenges, scientists are exploring innovative strategies to promote microbiome diversity and resilience, including the use of prebiotics, probiotics, and synbiotics – compounds that nourish and support the growth of beneficial microbes. Additionally, researchers are investigating the potential of microbiome-based therapies, such as fecal microbiota transplantation, which involves introducing a healthy microbial community into the gut to restore balance and combat disease.

As we embark on this journey into the microbial world within, we are faced with both challenges and opportunities. The microbiome represents a frontier of discovery, offering insights into the intricate interplay between our bodies and the microscopic organisms that call us home. By understanding and nurturing these microbial communities, we may unlock new pathways to promoting human health, preventing disease, and potentially reshaping the landscape of modern medicine.

This book invites you to explore the fascinating world of the microbiome, unraveling the complexities of these invisible ecosystems and their profound impact on our lives. Through a comprehensive exploration of the latest scientific discoveries, practical strategies, and emerging therapies, we aim to empower you with the knowledge and tools to cultivate a thriving microbiome, unlocking the full potential of these remarkable microbial allies for a healthier, more resilient future.

Chapter 1

Introduction: Journey Into the Microbial World Within

As I drifted off into a deep sleep one night, I never could have imagined the incredible dream journey I was about to embark on. I felt my consciousness plunging into a hidden realm, a veritable universe inside my body previously unknown. I was Alice heading down the rabbit hole, except this microscopic land was right inside my own gut.

With a sudden splash, I landed in a lush, thriving rainforest, only instead of majestic trees, there were long stalks of strange, unfamiliar flora undulating as far as my eyes could see. The air was alive with hundreds of bizarre creatures buzzing to and fro, floating past my face. I peered closer at one of these critters under my magnifying glass—it had spiraling, whip-like flagella (microscopic hair-like structures involved in the locomotion of a cell) and a bulbous oblong body, like some sort of single-celled space alien. In fact, I quickly realized that all these teeming "plants and animals were not macro-lifeforms at all—they were trillions of bacteria, viruses, fungi, and more that call the human body home. This was my microbiome. At that moment, a tiny voice arose from the underbrush—barely audible at first but rising in clarity and conviction.

It came from an old, wise bacteria named Bifido, who told me he had first colonized my gut days after birth. He had apparently been living there contentedly for decades without my knowledge. But today, he and his microbial brethren had an important message to share. "Welcome to your microbiome," squeaked Bifido. "We are the forgotten organ inside you—vital for your health, development, and even mood yet long ignored by science and medicine...until now. We have influenced your well-being since birth and will continue to do so for your entire life.

But the modern lifestyle is threatening microbial communities like never before, from industrialized food to over-sanitization to the overprescription of antibiotics. We must restore balance and harmony, or else humans and microbes will suffer grave consequences..." I awoke the next morning, still reeling from my vivid dream, yet feeling that it was a call to action. What did this all

mean? Who were the microbial denizens sharing my body, and what influence did they have? Why did this Bifido fellow seem so desperate for me to listen and understand? My mind swirling with questions,
I headed to my local library and devoured every microbiology textbook I could find. And slowly, astoundingly, the true importance of my journey into the microbial world began to take shape. The startling, almost inconceivable truth is that the human body contains roughly the same number of microbial cells as humans—a staggering 40 trillion bacteria, viruses, fungi, and other organisms inhabiting every nook and cranny of our inner and outer terrain. This bizarre kingdom had remained largely invisible essentially since the dawn of humankind.

But groundbreaking new discoveries in microbiology and scientific technology reveal these microbes do far more than simply live inside us: they perform vital functions that keep our immune defenses humming, brain functioning smoothly, organs operating in harmony, energy levels optimal, and more.

Our survival depends on hosting a flourishing, diverse community of microfauna inside and out. In fact, the emerging science of the microbiome reveals that we humans are actually more microbe than we are our own cells. Without these tiny passengers, we could not properly digest food, produce certain vitamins and nutrients, resist disease-causing germs, develop a robust immune army, detoxify pollutants, maintain a stable metabolism, and much more.

The 100 trillion microbes inhabiting a healthy adult human (several pounds worth!) contain a dizzying 100 times as many genes as our own genome. So, in essence, we are each an ambling colony composed predominantly of alien lifeforms! Far from being primitive germs, the microbes in and on us are actively shaping our physiology, behavior, immunity, appearance, and perhaps even mood and preferences daily. This winding jungle of bacteria, viruses, and more that I encountered in my dream is known as the "microbiome" (or "microbiota"), meaning simply "small life." Specifically, the human microbiome refers to the aggregate collection of trillions of organisms living symbiotically inside and on every person—mostly bacteria but also various yeasts, parasites, protozoa, worms, viruses, and more. Just as a bustling Amazon rainforest houses a wondrous diversity of flora and fauna, filling countless niches, so too do your inner terrain teams with thriving ecosystems of "friendly" microfauna serving important roles.

The number of unique bacterial strains living inside a healthy human bowel alone staggers belief estimated between 500 to 1000 different species. Microbial genes in the gut outnumber the 23,000 genes in the entire human genome 100-fold. Why such impressive complexity? Because over millions of co-evolutionary years between microbes and man, our microbial inhabitants have come to fulfill specialized roles—breaking down fiber, synthesizing certain vitamins, crowd-controlling pathogens, developing immune cells, and so much more. In return, they receive a consistent supply of nourishment and ideal habitat. Think of your own body as a mobile microbial metropolis! This cooperative arrangement between the human host and microbiota reaches back into prehistoric times. Over 60% of oral bacteria in modern man have recently been traced back to our hominid ancestors from millions of years past. For most of human existence, we lived close to an abundance of environmental bacteria, on nutritious whole foods and fermented fare. This allowed early groups to pass down diverse, resilient gut microbes over generations.

But in today's sterilized world of overly processed foods, hand sanitizers, and routine antibiotics, we've sabotaged what took eons of co-evolution to perfect. And we desperately need to re-discover our ancestral microbial roots if we hope to reverse the growing plague of 21st-century inflammatory disorders. Antimicrobial resistance is the term given by scientists to explain what is happening as our bodies grow more and more resistant to antibiotics – and scientists believe it is one of the greatest threats to human health in the coming years. While mankind had utilized yeasts for brewing and bacteria for fermenting various foods for ages, the scientific quest to first study and classify pathogenic microbes emerged in the 17th century with early pioneers such as Antonie van Leeuwenhoek peering into primitive microscopes to visualize these minuscule "animalcules" for the first time. Yet it would take another two centuries for the French chemist.

Louis Pasteur to conclusively link germs to infectious disease, ushering in antiseptic practices to keep patients sterile during surgery. This ignited the "war on germs' still raging today via hand sanitizers, antimicrobial cleansers, sterilizing wipes, and routine antibiotic prescriptions. Yet, in recent decades, researchers began questioning this hyper-vigilance against all bacteria. Could certain "healthy" microbes support bodily homeostasis rather than disrupt it? They discovered that specific organisms held anti-tumor potential, guarded against pathogens, enhanced digestion, stimulated immunity, and seemed intimately intertwined with health and wellbeing.

Thus, the revolutionary concept of "friendly bacteria" slowly emerged in niche scientific circles. But this notion of beneficial bugs was so astonishing that it barely penetrated mainstream medicine for decades. Finally, one blockbuster discovery after another in rapid succession over the past 20 years has thrust this once obscure topic firmly into the scientific mainstream, unleashing a Cambrian explosion of microbiome research. By sequencing all the genes within samples of human oral, skin, vaginal, and especially gut microbes, researchers illuminated entire ecosystems inside every person teeming with diverse organisms cooperating and communicating much like an actual rainforest. Moreover, these microscopic forests fell into consistent patterns—enterotype clusters dominated by select strains depending on individual diet, environment, hygiene, and health status.

In other words, just like in a macro habitat, the assortment of microbes in a given area depends on which environmental conditions allow certain species to thrive over others. By comparing robust, biodiverse microbiotas in traditional tribes with the severely depleted ones typical of Western nations, scientists awoke to the crucial importance these communities play in supporting our physiology—and just how gravely the modern lifestyle has endangered them. One researcher lamented, “We are running too low on microbes.”

While public awareness of this cutting-edge science remains generally low, make no mistake: decoding the human microbiome promises a sea change just as pivotal as the genetic revolution. It represents no less than an entirely new paradigm in medicine—one putting environment and lifestyle right alongside genes in determining health outcomes; scrutinizing how the inextricable links between host and microbiota impact the onset of most all chronic, inflammatory conditions today. Even the conservative National Institute of Health hails microbiome science as “one of the hottest new fields" that will soon open up promising probiotic therapies.

The remainder of our microbiome journey will survey all the latest electrifying discoveries in this burgeoning new field of biomedicine—from our gut all-stars impacting mood, immunity, and metabolism; to how vaginal microbes affect reproduction; and all about the newly discovered "skin microbiome" harboring billions of bacteria in diverse nooks and crannies. You will learn specifically how certain lifestyle choices badly disrupt delicate microbial communities, leading to systemic havoc and disease downstream; as well as how to nourish and cultivate flourishing, beneficial bug colonies through ancient dietary wisdom merged with modern science.

In essence, welcome to the world of "microbiomics" – perhaps the most overlooked key to radically transforming health in the 21st century and beyond merging ecology, nutrition, and medicine to uncover the microscopic foundations of vitality itself! Our microbial ancestors have been speaking to us in dreams and instincts: we must remember our primal roots, honor ancient partnerships forged over eons of coevolution, and restore lost bacterial harmonies. Within our trillions of humble microfauna may lurk answers to turn the tide against the epidemic of chronic illness sweeping post-industrial life while making "wellness for all" a biological reality instead of a buzzword. Are you ready to explore this journey into the uncharted wilderness within?

In an isolated research station deep within a steamy Borneo jungle, a curious team of scientists cluster around a flickering monitor, scarcely believing the revelations unfolding before their eyes. Peer over their shoulders, and you see the crystalline structure of a DNA double helix slowly rotating as if in zero gravity. Here's the shocker: it doesn't belong to *any* rainforest mammal or exotic bird indigenous to these parts.

Rather, it's teeming with the rich microbial genes extracted from...a routine stool sample! More commonly known as faeces! Yes, believe it or not, one pinch of organic bulk material taken directly from a volunteer's latrine contains 100-fold more wildly diverse genes than the entire human genome—thus offering a portal straight into the thriving bacterial jungle colonizing our inner terrain since birth. This "Poop Genome Project" represents one tiny puzzle piece gradually illuminating the sweeping mysteries of our "second genome"—that is, the dizzying array of microbial DNA interlaced with our own, shaping physiology and wellness at every turn even if biomedicine long ignored its existence. Widen the lens even further across various global efforts sequencing trillions of microbes occupying not just colons but also mouths, skin, and beyond, and you behold breakneck progress practically overnight towards decoding the human microbiome en masse—that bustling bacterial "rainforest" dwelling inside each of us.

Why such Herculean efforts to catalogue our inner ecosystems down to the last wriggling protozoa? Because burgeoning research ties nearly all chronic illness today ranging from autoimmunity and obesity right through anxiety and depression back to a common denominator: the withering of once lush and diverse microbiotas leaving dysfunctional, leaky terrain in their wake. The sobering consensus from scientists: We modern humans are "running too low" on microbial diversity; and our very survival now depends on replenishing what the post-industrial lifestyle has extinguished. But before detailing precisely how certain lifestyle missteps jeopardize delicate bug communities, first let's properly acquaint ourselves with the major microbial organs harboring distinctly unique balances of bacteria just waiting to be cultivated back to ancestral resilience! Make no mistake: the microbiome promises a total conceptual overhaul more pivotal than even decoding DNA itself.

But patient readers will emerge with both clarity on how to nourish their own robust, "rainforest-like" inner terrain; plus grasp the intricate links between microbe health and whole-body wellness for life. While trillions of bacteria blanketing our GI tract might dominate microbiome discussions (and rightfully so!), our mouths hardly qualify as microbial ghost towns either. In fact, some 700 varied species of bacteria alone make their home in diverse oral niches ranging from the hard, enamel-coated teeth to spongey, tissue-rich gums right through to the warm, moist environment coating the soft palate and tonsils. Why is such impressive diversity crammed inside an average-sized mouth? As with most microbiotas, form follows function.

Different oral microenvironments select for signature balances of bugs providing niche benefits—though often to the detriment of us human hosts! Let's break this down.For instance, the beneficial Streptococcus salivarius strain K12 is lauded for its stellar ability to crowd out pathogens, dissolve tooth-corroding acids and dampen inflammation in gum tissue.

Yet its distant cousin S. mutans oozes onto pearly white teeth bathed in sugary films, fermenting sucrose into sticky biofilms bristling with cavity-forming acid. Meanwhile, Porphyromonas gingivalis burrows deep into tender gum pockets filled with bleeding, irritated tissue—there unleashing its true nefarious aim to trigger systemic inflammation linked to everything from heart disease right through to dementia and cancer! Clearly, as leading microbe ecologist Prof. Stanley Falkow laments, "Bacteria are not all created equal"—and nowhere is it more apparent than our oral cavity where a delicate, complex balance between so-called "good bugs" and "bad bugs" make it either a smooth gateway for whole body health and immunity...or alternatively, an underappreciated epicentre of chronic inflammation stealthily undermining wellness over decades. And while aggressive periodontal pathogens like that miscreant P. gingivalis often dominate headlines for wreaking dental havoc, even more alarming is the growing roster of usually innocuous strains that turn rogue when microbial harmony falters amid the modern lifestyle. Although 95% of oral bacteria do not irritate under balanced conditions,

Cocci sapiens bacteria that harmlessly inhabit many mouths can trigger painful abscessed infections once immunity falters. Similarly, the slow conversion of once-innocuous strep into the tooth-rotting S. mutans typifies how soaring starch and sugar intake coupled with inadequate nourishing whole foods feeds chaotic pathogenic overgrowth at the expense of gentle but vital base strains.

This is why dentists and doctors urge a healthy, balanced diet with ideally little sugar intake. Even common Staphylococcus hominis strep, while helpful for fermenting fungi in the healthy polymicrobial landscape, grows aggressively in the absence of rivals—spurring inflammation. In essence, the take-home is that no single "bad bacteria" exists—rather, it is the wholesale unraveling of proportions between "good" and "bad" bacteria that so profoundly impacts oral immunity and inflammatory cascade. And while dentists today primarily focus their artillery on neutralizing select pathogens already running rampant, researchers now realize the ultimate answer lies in proactively promoting the balanced, thriving oral "rainforests" that keep harmful strains naturally curbed year after year. So critical is oral ecology that babies born vaginally receive vital foundational inocula from pregnant persons preparing the baby's mouth for long-term harmony. In recent years, we have all heard much about the gut and its crucial role in our bodies. The gut predictably hogs the microbiome limelight, given its indispensable role in promoting digestion, immunity and even regulating appetite, mood, and metabolism.

Yet far less glamorous sites ranging from your armpits to the moist abyss between toes also qualify as bustling bacterial biomes finely tuned after eons of coevolution. Of course, "skin microbiome" hardly conjures images of a thriving rainforest so much as a rather unsavory swamp, but make no mistake—those billions of bacteria coating every inch of your outer terrain are equally vital sentinels, both crowding out invasive pathogens and even "educating" the underlying immune cells.

Researchers today estimate a staggering 1000 or more bacterial species tending the average arm, leg, chest, and back—forming diverse metropolises based on moisture, texture, hair follicle density, and oil production unique to each "locale." Only here, the balancing act tilts more towards oily, anaerobic Corynebacterium; Propioni- and Staphylococcus predominate. Interestingly, while puberty and hormonal swings trigger microbiome-wide shifts, adolescent skin endures the biggest upheaval as formerly dominant Staphylococcus dip levels to allow lipophilic Propionibacterium, the upper hand in oil-rich eruptions of acne and blackheads. I think most of us had bad skin or breakouts in our teens! Later in life, skin aging markers positively correlate with shrinking microbial diversity, allowing normally minority pathobionts to stake claims in plots of drying, and sagging dermis, leading to irritation.

What's even more crucial than promoting an event, robust skin terrain? Those colonies serve as border patrol to deeper lymphoid organs, educating naive immune cells about which bacteria require attack versus tolerance to maintain harmony. Hence why, deliberately damaging this barrier through harsh soaps or antimicrobial creams alone fuels allergenic, autoimmune, and inflammatory disorders downstream as the body loses the ability to distinguish friend from foe.

Conversely, supporting microbiome health through gentle cleansing, prebiotic skin foods and homeostasis-enhancing modalities like infrared sauna, dry brushing, and earthing grounds may all assist proper tolerance signaling. This is why you should be very careful with potentially damaging skin routines, whether surgical (like Botox), or chemical peels. But what is the connection between the skin and the gut?

While each human body habitat teams with signature microbiotas tailored to environmental pressures, no site interacts more intimately with your thoughts, cravings, and emotions than the great gastrointestinal tract—home to two dense pounds of bacteria engulfing the stomach, small intestine and especially a large, distended colon positively teeming with hungry bug residents awaiting their next "delivery" to jumpstart fermentation and energy production!

Unsurprisingly, then, science confirms direct communication lines between gut bugs and the brain through the so-called gut-brain axis determining much of mood, motivation, and even food selection! Via direct neural networks plus circulating endocrine hormones and inflammatory cytokines, intestinal flora engineer responses promoting their survival—be it through ramping appetite when favorite prebiotic fibers run scarce or alternatively calming stress reactions that can destabilize sensitive microecology. Conversely, unhealthy diets high in inflammatory foods coupled with routine antibiotics decimate once-balanced intestinal tracts, allowing opportunists to upregulate signals stoking anxiety and gloom while killing inspiration to eat wholesome, nurturing fare.

Research spotlighting gut-brain crosstalk remains in relative infancy, but already psychobiotics and other interventions fine-tuning microbial balance demonstrate striking potential to alleviate symptoms of major depressive and bipolar disorders, autism spectrum, and perhaps dozens more afflictions rooted in systemic inflammation. One research director noted: "We have created an environment in which microbes can make us fat." Hence, there is a growing plea from thought leaders to view humans not as autonomous entities but as elaborate patchworks of human and microbial metabolism intimately intertwined through eons of coevolution.

Long considered sterile compared to more bustling body sites, breakthrough research now illuminates the vaginal canal as a vibrant though delicately balanced microbiome in its own right—one profoundly impacting reproductive outcomes. Researchers today estimate more than 270 strains of protective Lactobacillus species form lactic acid strongholds, warding off invasive yeast and pathogenic bacteria in most women's lower crevices.

However, pregnancy, hormonal birth control, antibiotics, and certainly the growing scourge of douching with irritating concoctions alters chemistry enough to allow outside opportunists like Gardnerella vaginalis or Candida fungal strains to stake strongholds, instigating the all-too-common inflammation, irritation, and discharge emblematic of bacterial vaginosis and yeast infections.

But aside from provoking discomfort in the moment, these conditions that compromise microbial harmony also sabotage fertility and amplify the risk for ascending infections post-birth. Why such dire results when a delicate balance falters? Some experts believe the initial seeding of Lactobacillus strains from mother to newborn during vaginal birth (just like in the oral cavity) perpetuates transgenerational stability.

Hence, C-section delivery precludes this colonization process, allowing outside opportunists easier access. However, deliberate supplementation of tactics through suppository, oral probiotics, or bio-matched transplants shows particular promise in restoring homeostasis, especially when combined with eliminating sugary foods, alcohol, and irritating feminine hygiene products from the daily lineup while opting for breathable cotton undergarments.

Step back from this whirlwind review touring your body's complex bacterial terrain in all its diversity, and an undeniable reality crystallizes into view: far more foreign cells thrive inside us humans than actual human parts! By the most conservative estimates, microbes outnumber our total tally of 30 trillion-odd human cells by at least 4-to-1; though by some accounts, their advantage soars closer to 10-to-1.

From abundant energy-harvesting Firmicutes and inflammation-regulating Bacteroides blankets enfolding our lower intestines through hardy Actinobacteria dominating moist folds and crannies of our epidermis down into equally specialized communities dotting the urogenital tract—we humans qualify as bustling microbial metropolises as much as autonomous lifeforms.

This radical reformulation should give us some perspective, revealing Homosapiens more akin to ambling colonies rather than independent entities. Yet it simultaneously empowers each body host with the humbling knowledge that lifestyle choices, not prewritten genetic fate, remain the ultimate arbiters of destiny. Just as habitats ranging from lush rainforest to desolate tundra select for signature flora and fauna, so too our everyday decisions on diet, medications, hygiene and more determine if diverse beneficial commensal strains flourish over hazardous pathobionts.

Chapter 2
The Complex Communities of Microbes Within Us

We hear a lot about the human body, but it's essential to understand what we are made up of. The human body comprises over 30 trillion human cells, but scientists have discovered we harbor a far vaster population of tiny microbes dwelling inside us.

Over 100 trillion diverse bacteria, viruses, fungi, and other microorganisms inhabit spaces on our skin, gut, and organs and tissues. Just as the rainforest or coral reef depends on the interplay of thousands of plant and animal species, we rely on a thriving community of "friendly" microbes to help run critical bodily processes. What do all these microscopic bugs do for us exactly? More than we ever imagined, as it turns out. Researchers comparing germ-free mice raised in sterile bubbles to normal mice were stunned to find the microbe-free rodents had defective immune systems and trouble digesting food without their resident microbes. Over 90% of the cells in our body are not even human - they belong to the fungi, bacteria, and viruses that sculpt our inner ecosystem.

It's kind of disgusting to think about, but all humans have microscopic bacteria that live on our skin. Far from being freeloaders, our microbial communities play indispensable roles in regulating immunity, aiding digestion, manufacturing vitamins, protecting from pathogens, and supporting healthy mood and metabolism.

The collection of microorganisms in and on our bodies is termed the "human microbiome." Each person's community of bacteria alone contains over 1000 diverse species, with the gastrointestinal tract hosting the densest populations. The number of genes present across the human microbiome is 150 times greater than in the human genome. This means that as our human cells work together, our microbes also form interconnected communities that profoundly shape our biology.

The Human Microbiome Project was launched in 2008 to catalog and map this forgotten organ inside us. With stool samples from hundreds of volunteers, researchers were stunned that healthy people shared less than 40% of bacterial strains. This signals that each person's microbiome is as unique as their fingerprint. Yet despite this diversity, our microbial companions tend to carry out similar essential duties - the specifics of who does what vary between individuals.

Like band members with different hairstyles and outfits, the microbes fill characteristic roles, but the performers themselves differ. It turns out that the route of childbirth, method of infant feeding, medication use, diet, lifestyle habits, and other environmental exposures shape the assortment of microbes acquired. Once established, many of our microscopic partners stick with us for decades, although periodic disturbances can upset the balance. That's why preserving microbiome diversity and resilience is vital for withstanding stressors and avoiding illness.

But before diving into how we can tend our gardens of invisible bacteria, let's peek first at who these microbes are and what they normally do inside us. Bacteria make up most of the members of the standard human microbiome. The most prolific phyla in our gastrointestinal tract include Bacteroidetes, Firmicutes, Actinobacteria, and Proteobacteria. The bacteria within our large intestine are anaerobes, meaning they thrive without oxygen by fermenting carbohydrates from our diet.

This produces inflammation-lowering short-chain fatty acids that nourish our gut lining and benefit whole-body metabolism. Other well-studied genera living in the human intestine include Lactobacillus, Bifidobacterium, Escherichia and Clostridium species. Researchers have discovered that the balance of these bacteria tribes helps regulate digestion, immunity, and mood. For example, lower levels of fecal bacteria and other anti-inflammatory species have been tied to depression and irritable bowel disorders. This reveals the importance of preserving microbiome harmony.

Beyond bacteria, the human virome consists of eukaryotic fungi and viruses that inhabit us. The mycobiome refers to the community of fungi and yeasts dwelling in areas like our mouth, skin, and genitals. Species of candida yeast help digest carbohydrates, while Malassezia breaks down oils on our scalp. An imbalance favoring pathogenic fungi is linked to oral thrush and stubborn skin or nail infections. This is why infections like these are often treated with antifungal medications, which restore this imbalance. The human virome consists of beneficial bacteriophages and viruses infecting our microbiome. Bacteriophages are viruses that insert into bacteria to replicate. Researchers suggest bacteriophages help regulate bacterial densities and gene transfer. An estimated 90% of bacteria harbor at least one virus.

The virome can become imbalanced in conditions like inflammatory bowel disease, indicating the importance of harmony between viruses and bacteria within us. The assortment of microbes living on and inside humans have specially evolved to partner with us by filling metabolic and immune support roles our human cells cannot perform alone. Our microbes are so vital that without them, we would surely perish. In return for room and board, our microbial communities help digest our food, supply us with vitamins, tune our immunity, protect us from invading pathogens, nurture the lining of our gut, and even produce molecules that regulate our mood and behavior. It turns out we have outsourced many basic bodily processes to the members of our microbiome.

For example, we lack the enzymes required to digest tough plant polysaccharides from the vegetables, grains, and legumes we eat. Fortunately, our gut bacteria produce carbohydrate-active enzymes capable of unlocking these nutrients for us. The short-chain fatty acids created from their fermentation nourish our colonocytes to support gut barrier integrity. Amazingly, over 60% of the metabolites in our bloodstream are microbial in origin - not human! This illustrates how profoundly bacteria underpin our biochemistry. Another vital role played by our resident microbes relates to education and the balancing of our immune system. Exposure to environmental microbes from birth helps train our monocytes, macrophages, and T-lymphocytes to respond appropriately to pathogens. Meanwhile specific bacterial metabolites help immune cells differentiate between harmless and harmful triggers to prevent overactive inflammatory responses.

Researchers have shown that children exposed to more diverse milk, soil, and plant microbes have reduced allergy and asthma rates later in life thanks to this immune-balancing effect. This may also explain why having a house that's overly clean or sterile is pointless – sometimes, we need to be exposed to a small amount of bacteria to make us immune to certain harmful bacteria. Incredibly, research over the past decade has revealed that the influence of our microbiome extends far beyond immunity and digestion to impact cardiovascular health, liver function, metabolism, hormones that regulate appetite, and even our moods and cognition. Germ-free mice display altered stress responses and scientists have identified a gut-brain axis linking our resident microbes to mental health. Even obesity has been tied to specific ratios of Bacteroides, firmicutes, and other bacterial tribes living in our intestines.

This means the ballet of invisible organisms within us is involved in far more than we ever presumed. Each section of our digestive tract provides specialized conditions suitable for signature communities of microbes. Our mouth harbors over 700 species of bacteria, which begin breaking down starches. Skin microbiota helps form a protective barrier against pathogens. The vaginal microbiome plays a key role in preventing the growth of yeast and pathogenic strains. Our intestinal ecosystems contain thousands of bacteria, viruses, and fungi supporting metabolic, digestive, and detox functions. All of them work in incredible synchronicity towards nourishing our human cells.

Now that we've glimpsed the assortment of microbial companions inhabiting our skin, mouth, lungs, and digestive system, let's peek at what they typically do for us day-to-day to keep our bodily ecosystem humming. The bacteria in our mouth are among the first digesters that break down leftover bits of carbohydrates after we finish a meal. Saliva supports strains of strep and lactobacilli that convert residual starches into short-chain fatty acids and smaller sugars. They also help form dental plaque, protect our enamel, and produce compounds called bacteriocins,

which hinder oral pathogens. Beyond digestion, some oral bacteria may stimulate inflammation, contributing to periodontitis if their numbers dominate. This reveals the importance of microbial balance. Traveling downwards from our mouth, we reach the longest portion of our digestive tract - the small intestine. Villi projecting into the lumen increases surface area for absorption of liberated nutrients. Here we find fewer microbes but vital ones that continue dismantling carbs, proteins, and fats from our meals. Lactobacilli, streptococci and bacilli in upper small intestinal niches transform amino acids and fats while reinforcing tight junctions between epithelial cells.

The ileum's lowered pH and rebellion fractions permit acid-loving lactobacilli, actinomycetes, and fungi to dwell. These microbes cooperate with nearby human cells to absorb dietary iron, magnesium, calcium, zinc, and phytochemicals. Further along, the large intestine represents the most densely populated microbial niche, containing up to 70% of all the bacteria in our body. Trillions of anaerobic microorganisms make their home in our colon and rectum, where sluggish transit permits the fermentation of residual nutrients. Bacteroides, bifidobacteria, clostridia, and other tribes feast upon undigested carbohydrates such as pectins, gums, inulins, and resistant starches that escaped upstream absorption. The resulting short-chain fatty acids, including acetate, propionate, and butyrate, play vital signaling and energy roles.

After enzymatic breakdown, our colonocytes absorb these bacterial metabolites for liver processing. Butyrate, in particular, nourishes the epithelial cells lining our colon to reinforce gut barrier integrity.

This helps seal gaps between enterocytes that could permit entry of inflammatory compounds, allergens, or pathogenic bacteria. Short-chain fatty acids also favor community richness among symbionts while suppressing colonization by detrimental blooms of candida or clostridia species. Bacteriocins and organic acids secreted by our microbiome similarly hinder the overgrowth of opportunistic microbes that harm health. This illustrates how bacteria self-regulate. Also beginning in the colon are phases II and III of bile acid transformation. Our liver initially produces cholic acid from cholesterol and then pumps it into the small intestine to emulsify fats. Yet over 95% gets reabsorbed and transported back to the liver for reuse. Gut bacteria unlock a small fraction of our bile, transforming primary acids into secondary ones like lithocholic and deoxycholic derivatives. This liberates antioxidant compounds that regulate fat and glucose metabolism.

The bacteria help recycle our bile, and not having enough of these bacteria in your body could explain why someone might suffer from heartburn or indigestion. Lower in the large intestine, final enzymatic catalysis by hostile bacteria splits apart the remaining carbohydrate fractions. Methane and hydrogen produced by these final fermentation steps are then expelled as flatus. Though the aroma of such microbial mechanics may embarrass us, these activities illustrate the teamwork between Gastroenterol, hepatic, and microbial metabolic cycles that sustain our life and health. This cooperation between human and microbial metabolism allows us to extract a bounty of calories and nutrients from various phytochemicals found in grains, produce, herbs, and fermented foods. Humans specifically evolved to rely upon gut bacteria to unlock these substrates.

Their fermentation byproducts signal our genes to properly develop, digest, absorb, detoxify, and repair. Without this microbial backbone, human life would definitely not exist. But what is the link between our microbiome and vitamins? Incredibly, our microbiome provides roughly 9 of the 13 vitamins we require, including biotin, cobalamin, folates, nicotinic acid, pantothenic acid, pyridoxine, riboflavin, and thiamin! Germ-free animals need to be supplemented with these essential compounds, proving humans did not evolve to synthesize them alone. Beyond basic nutrition, gut bacteria partake in the transformation of plant compounds into bioactive metabolites that regulate immunity and hormones. Phenolics in green tea transformed by our microbiome aid blood sugar regulation.

The microbial alchemy deep in our guts literally shapes human health. This begins with the profound inter-kingdom communication between bacterial, human, and even fungal genes within us. Researchers have discovered that metabolites and messaging molecules secreted by our microbiome act as hormones coordinating our physiology. These compounds switch human genes on or off to appropriately develop digestive capacity, immune tolerance, and metabolic regulation in infancy.

They fine-tune nutrient absorption rates later on. Even more extraordinarily, they modulate appetite by triggering intestinal satiety hormones directly linked to weight and insulin sensitivity regulation. Such microbial signaling compounds include metabolites like butyrate, secondary bile acids, indoles, propionate, and polysaccharide A, as well as biogenic elements like LPS, peptides, and gases that trigger immune reactions. By excreting neurotransmitter-like compounds, including acetylcholine, catecholamines, histamine, melatonin, and serotonin, our gut microbes regulate mood, motivation, focus, and sleep-wake cycles. Even diseases like autism, Parkinson's, and depression correlate with an imbalance of such microbial messaging molecules.

This spotlights how deeply our bacteria influence human health. Incredibly, mothers impart foundational microbes containing memory immune cells and metabolites to their infants during childbirth and breastfeeding. These pioneer species are believed to help appropriately develop everything from our gut lining to our blood-brain barrier. They educate our naive monocytes while secreting compounds that upregulate tight junction proteins, sealing gaps between intestinal epithelial cells. Such bioactive factors promote angiogenesis and trigger mucosal immunity. Many originate from an infant's first gastrointestinal bacteria. Together, this illustrates how the assortment of microbes acquired from mom, food, and environment in infancy influences far more than just digestion and immunity. Because humans co-evolved relying on microbial biochemistry and signaling, every organ system, ranging from our guts to our brains, depends upon balanced microbiome communities.

Research reveals disruptions contribute to disorders ranging from depression, obesity, and diabetes to autoimmunity, allergy, and even neurocognitive conditions. Therefore, nurturing microbiome diversity has implications for whole-body health. In summary, the human microbiome encompasses highly adapted microbial ecosystems inhabiting our skin, mouth, lungs, and digestive system. These communities contain trillions of bacteria, viruses, and fungi that help digest our food, supply vitamins and nutrients, nurture organs, fine-tune metabolic health, support detoxification, reinforce our gut lining, regulate immunity, and even moderate neurotransmitters that shape our mood and behavior. In return we provide them with a steady room and board. We have evolved intimate interdependence upon these microorganisms for regulating basic physiology.

Disruptions in their community structure, signaling, and metabolites contribute to rising lifestyle diseases. Therefore, maintaining a balanced, thriving microbiome through proper feeding and protection is the key to lifelong health and well-being.

Chapter 3
How Microbes Influence Human Health and Disease

The human body carries within it a vast, complex, and dynamic microscopic world - a hidden ecosystem of microbial organisms, including bacteria, viruses, fungi, and protozoa.

While the germ theory rightly made us wary of pathogenic microbes, researchers have discovered that most microbes within and upon our bodies are harmless or beneficial.

An exploding array of research continues to uncover how this microbiome - the genetic material of all these microbes within us - is intimately linked to human health and disease. As we saw earlier,the human microbiome contains over 100 trillion microbial cells, outnumbering human body cells by an estimated 10 to 1. As our understanding grows, it becomes apparent that these previously ignored internal ecosystems play indispensable roles in training our immune systems, crowding out pathogens, extracting nutrients and energy from our diets, producing vital compounds like vitamins and anti-inflammatories, and even influencing our behavior and moods.

Dysbiosis (disruptions to the balance of these communities) is now implicated in a wide spectrum of conditions and illnesses, including autoimmune disorders, allergies, obesity, malnutrition, diabetes, autism, anxiety, depression, and more.

While research into the profound impacts of the human microbiome blossomed this century, the concept of health-giving gut flora was intuitive to ancient cultures. Fermented foods and beverages like kimchi, kefir, and kombucha were revered for their healing properties across traditions from Ancient China and India to Pre-Columbian Americas. Roman soldiers carried fermented wine to stay vigorous on long campaigns.
Let's start at the beginning of human life. Babies are born essentially sterile, with no microbes lining their pristine gastrointestinal tracts, skin, or other body habitats. This is why newly-born babies are so vulnerable to any type of bacteria. Yet within hours, a cascading microbial colonization begins to profoundly shape future health and disease outcomes.

Pioneering neonatology microbiome research is shedding light on this formative process, raising the stakes for getting it right. Vaginally delivered newborns pick up protective maternal microbes like Lactobacillus and Prevotella as they pass through the birth canal. Breastfeeding furnishes special oligosaccharides to feed these early gut colonizers, plus continued inoculations of key strains like Bifidobacteria. Meanwhile skin, mouth, and lung communities assemble through contact. By age 3, the young microbiome stabilizes into an adult-like configuration marked by high levels of health-promoting microbial diversity.

However, there is no conclusive research to suggest that babies delivered by C-section or babies that are not breastfed will miss out on key nutrients. Early antibiotic exposures can decimate vulnerable microbiomes. Such dysbiosis may leave lasting imprints on metabolism and immunity for decades. However, techniques like swabbing C-section newborns with vaginal fluids, nursing longer, and avoiding unnecessary antibiotics increasingly offer means to restore microbial richness. The future of medicine may focus as much on optimal seeding of our microbiomes at birth as any factor under human control.

While microbes dot everybody's habitat, from skin to lungs to genital tract, the intestinal microbiome remains the most crucial hub for human health - home to over 70% of our resident bacteria. Numbering in the tens of trillions and weighing up to 5 pounds, these gut microbes interface with key systems regulating immunity, metabolism, hormones, nerve signals, and mental health. Imbalances in gut microbiota are now linked to dozens of afflictions, from inflammatory bowel disease and metabolic syndrome to anxiety, MS, and autism. However, it is tough as of yet to verify why these developments may occur. Compositionally, our gut microbiomes resemble lush rainforests filled with hundreds of interdependent species.

While a core set of common phyla dominate, including Bacteroides, Firmicutes, and Actinobacteria, ratios vary dramatically from person to person, influenced by genes, age, diet, environment, and more. Two people may share just 10% of their gut species in common. Diversity provides functional resilience - like an economy with many industries. When key groups falter, the disease can gain footing. While links between disease and specific pathogens emerged thanks to germ theory, diagnoses now spotlight broader communities. For example IBD develops not from one bug but collapsing microbial diversity.

Oral Verrucomicrobia crowds out pneumonia-causing species. Such findings are spawning a new science of cultivating hardy, biodiverse guts via probiotics, prebiotics, and healthful living - not waging war on single-agent foes. For systemic balance search and destroy may prove the best path to microbial health. Microbiome dialogue with human cells remains lively but opaque. Gut microbes send signals via metabolites like short-chain fatty acids, neurotransmitters including over 50% of natural dopamine and even quorum-sensing compounds. Meanwhile, our human cells reply through skin secretions called mucins, antibodies, and signaling proteins called cytokines. Researchers have identified over 15 signalling routes between microbes and humans, which is surely the tip of the iceberg. As communication networks come into focus, so may our powers to hack it. While the intestinal lining must smile openly to absorb nutrients, it relies on the immune system to sound alarms when pathogens approach.

Gut microbes play indispensable roles in training and coordinating this defence - signalling reaction urgency, crowding out pathogens, and calming overly aggressive reactions. Dysbiosis that derails such immune dialogue opens the door to myriad common afflictions centered around inflammation. Inflammation forms a cornerstone of human immunity and healing - marshaling cellular defenders against injury and infection.

But improperly balanced or prolonged inflammation also plays a role in virtually every modern scourge, from autoimmunity to heart disease. Intestinal microbes train developing immune cells throughout life, teaching what to attack and what to tolerate. Without such education, our defenses can become hazardously triggering - attacking benign targets like pollens or peanuts while failing to focus fire on true pathogens.

Pro-inflammatory species spur immune reactions, triggering inflammation, while tolerant microbes signal restraint, neutralizing alarmist signals. Resetting this dialogue may offer paths to resolving countless mysteries, like why some people develop allergies or autoimmunity. Does microbial miscommunication confuse host cells into friendly fire assaults? An ability to select and amplify microbial peacemakers could transform medicine - such abilities may be within reach. Consider promising trials where IBD patients received healthy donor stool to restock tolerant species and quiet gut inflammation. By better understanding the languages of microbe-immune chemical lingo, we may learn to converse directly with it.

With over one-third of adults now obese, decoding links between the human microbiome and metabolism has become an urgent priority. While feasting and inactivity feed the crisis, our resident microbes also play central roles in influencing what we digest, signaling satiety, and coordinating whole-body metabolism.

Improving dietary behaviors and exercise often fail long-term without addressing dysfunctional gut ecosystems. Western microbiomes differ markedly from lean rural dwellers and pre-industrial humans - less diverse and dominated by Firmicutes over Bacteroidetes. Animal models confirm lean versus obese microbiomes alone induce their divergent phenotypes on receipt. Even without overfeeding, overweight microbes can drive metabolic syndrome.

This undergirds the exciting potential for microbial transplants from healthy donors to override unhealthy programming. Gut microbes metabolize compounds human enzymes cannot unlock from plant foods - extracting ~10% extra calories per gram.

They also generate vital micronutrients like vitamin K. Key genera boost nutrient absorption by enhancing gut barrier integrity, calming inflammation, and improving insulin response. Dysbiosis allows energy harvest and inflammation to continue unchecked, driving obesity. Plus, it also needs to be mentioned that we lead far more sedentary lifestyles than previous generations did. Some opportunists like Hungatella even directly trigger sugar cravings and habituate hosts to high-fat diets. Identifying and sidelining such saboteurs offers promising avenues for weight loss aids. As our diets and lifestyles rapidly change, resident microbes fail to keep pace. Discordance between past Evolutionary conditions (under which our microbiomes evolved) and modern settings continues spawning Western epidemics of obesity, allergy, and autoimmunity. Some prospect that future prebiotic diets paired with probiotic biohacking may help update our internal ecosystems to thrive amidst modern realities. For now, obesity remains microbiome mismatch made flesh - emblematic of the power of tiny organisms over human health.

Chapter 4

The Gut-Brain Axis: Microbes and Mood

The digestive tract is often viewed as carrying out a simple, mechanical function in the body - breakdown of food to extract nutrients and energy. However, we now understand it serves a far more complex role, communicating back and forth with the brain through nerves, hormones and microbial molecules in a vast, interconnected network.

This gut-brain signalling highway enables trillions of microbes residing in our intestines to talk to our central nervous system, influencing everything from mood and stress levels to food cravings and sleep patterns. Emerging research indicates that supporting the diversity and health of our gut microbes could be key to overcoming anxiety, depression, fatigue and other common mental health issues. These microscopic organisms produce many of the same neurotransmitters as our own neurons, allowing them to profoundly shape brain processes.

They also direct the function of the vagus nerve, the main channel of communication between gut and mind. When certain species disappear or become overgrown, we lose this beneficial microbial influence, contributing to inflammation, immune imbalances and 'leaky gut' that can manifest as psychological troubles. Resetting this two-way dialogue by nourishing our good bacteria promises exciting possibilities for treating stubborn mood disorders, attentional difficulties, obsessive behaviors and even neurodegenerative disease. The gut-brain connection gives new meaning to the old adage that "you are what you eat", suggesting our daily dietary and lifestyle choices lay the foundations for mental wellbeing. Emerging studies also indicate transmission of certain microbes from mother to infant may help stimulate development of the social brain. Understanding these direct pathways between microbe and mind will allow us to intentionally cultivate internal ecosystems that foster psychological resilience. At the heart of the gut-brain axis lies the vagus nerve, the main highway of information flow between digestive tract and central nervous system.

This critical channel consists of 100 million nerve fibers that conveys sensory signals about the state of our inner ecosystem up to the brain, while delivering motor commands down from central control centers. As the longest of the cranial nerves, the vagus nerve meanders its way from the brainstem through the neck, chest and abdomen, interfacing with major organs along the way. This bustling neural motorway enables a constant, lively dialogue between microbes in the gut and regions of the brain pivotal for psychosocial functioning like the hypothalamus, amygdala, and prefrontal cortex. Signals originating in the intestines keep the brain updated about what's occurring downstream, while the brain tightly regulates digestive processes and intestinal permeability.

However, chronic imbalance in the gut microenvironment can lead to aberrant sensory signalling to the central nervous system along this conduit, driving inflammation that underlies 'gut-brain' mood disorders. Imbalances in the gut microbiome, what we now know is dysbiosis, directly impact signaling along the gut-brain highway, contributing to rising rates of mood disorders worldwide. As beneficial microbes dwindle and inflammatory species bloom, they produce fewer happy-brain chemicals like serotonin and GABA while churning out stress chemicals that prevent us from feeling calm, satisfied, and optimistic. One study found major depressive disorder patients had up to 90% less gut Faecalibacterium species that produce the feel-good molecule butyrate. Without adequate daily nourishment from veggies, fibers, and fermented foods, our good bacteria starve, leaving space for opportunists to take over. I'm sure we all know someone who regularly eats natural yogurt or Kombucha and loves talking about how it makes their gut feel good! A lack of fermented foods allows lipopolysaccharides and other inflammatory compounds from pathogenic strains to leak from damaged intestinal lining into circulation.

Once in the bloodstream, these rogue microbial agents spark low-grade inflammation and immune activation implicated in anxiety, OCD, Seasonal Affective Disorder, and Parkinson's disease progression. Research indicates overgrowth of Candida yeast in the small intestine drives brain fog, fatigue, and mood swings by releasing toxins that short-circuit communication along the vagus nerve. Studies also link altered Firmicutes to Bacteroidetes ratios characteristic of dysbiosis to depressive symptoms and lower quality of life scores.

Together, these findings confirm unhealthy shifts in our microbiome lay the foundations for inflammation-related psychological issues. However, in one study, depressed individuals who consumed a daily probiotic saw significant improvement in mood scores after just three weeks, highlighting the rapid influence microbes wield over mental health. Strategies like dietary prebiotics, microbial transfers, and psychobiotics may effectively 'debug' the gut-brain axis by flooding our intestines with stable, resilience-promoting species. Ultimately, shedding light on these direct conduits between microbes and the mind will allow clinicians to intentionally seed inner ecosystems that foster psychological wellness from the inside out.

While the bidirectional pathways enabling gut flora to shape anxiety and depression were only recently discovered, traditional Eastern medicinal systems recognized links between the intestines and emotional health centuries ago. The traditional Chinese concept of the "microbiome-gut-brain axis" viewed the digestive tract as inextricably tied to mental processes and consciousness via continual cellular signalling.

Within this framework, chronic GI issues and microbial imbalance laid the pathway for eventual mood disturbance by sparking inflammation that clouded the mind. Free flow of energy and information throughout the microbiome-gut-brain network was considered essential for psychological harmony and clarity of thought. Bioactive compounds in tonic herbs and fermented foods were believed to nourish this communication circuitry, preventing mental distress. Modern investigations now confirm many precepts of this ancient model of gut-brain interaction, shedding light on how disruptions in microbial signals along the vagus highway manifest as psychological distress.

Decreased microbial richness, surface adhesion and carb metabolism provoke anxiety in animal models, while oral administration of Bifidobacterium and Lactobacillus strains mitigates these behaviors. Human studies reveal probiotic supplementation in healthy adults enhances brain connectivity in regions that process sensory information and quiet anxiety. Together, these findings validate the notion of an interconnected ecosystem spanning from microbe to mind that breaks down when critical species are lost.

In addition to impacting stress reactivity and mood regulation, emerging findings indicate the gut microbiome even sculpts complex social behaviors in mammals by secreting compounds that nourish neurons. Animal studies report decreased sociability and social reward circuit activation in germ-free subjects. These social deficiencies were reversed by adding certain Acinetobacter and Lactobacillus strains known to produce the precursor for the feel-good brain chemical serotonin. Oxytocin and vasopressin are keystone neuropeptides for promoting social bonding, trust and cooperative behaviors in mammals. University of Cork scientists in Ireland recently discovered a gut microbe species called Lactobacillus reuteri boosts oxytocin levels through the all-important vagus nerve, enhancing social cognition in rodent models. This reveals direct molecular conduits linking microbial members to neuronal networks governing social conduct. It may be no coincidence that conditions marked by social difficulties like ASD are frequently accompanied by GI issues and dysbiosis.

Transplanting faecal microbes from high functioning autistic kids into mice provokes hallmark symptoms like anti-social tendencies and repetitive behaviors. Together these revelations confirm Freud's famous quote that "the ego is first and foremost a bodily ego" arising from interoceptive signals about the internal biological milieu. In this view, an unstable, inflamed gut environment breeds psychological turmoil that hampers social thriving. So, how does our “social brain” develop from birth? In the first years of life, the developing 'social brain' grows exponentially, laying the foundations for emotional health.

During this time, one of the most influential inputs sculpting neonatal neurobiology is the mother's microbiome passed down through vaginal birth and breast milk. These seeded strains colonize the infant gut, where bacterial chemicals support neuronal connections and neurotransmitter supply vital for future sociability. For example, researchers at Baylor College discovered that vaginally-born baboons, who received key microbes like Prevotella and Lactobacillus, demonstrated greater social engagement and curiosity later on compared to C-section-delivered monkeys. This microbial transfer from mother to baby stimulates genes pivotal for oxytocin signaling, priming complex social behaviors. Later dysbiosis likely erodes these critical neural networks governing bonding and cooperation. Similarly, infants fed exclusively breast milk reap lifelong emotional intelligence benefits from the 770 strains of bacteria uniquely found in human milk. Researchers propose the high dopamine levels needed for peak social development depend on nutrient metabolites and tryptophan provided by this "psychobiotic" breastmilk community.

This highlights the need for mothers to nurture their own microbiome balance when conceiving and nursing actively. In this way, reproductive ecology transmits psychic resilience intergenerationally through microbes, contrasting the individualistic genetic model of mental illness. Appreciating the interconnected nature of the family microbiome calls us to a collective commitment to seeding future generations - both inner and outer - with the commensal species that nourish psychological harmony.

Chapter 5

Microbiome Imbalances and Chronic Illness

An imbalance in our microbiome, that bustling community of microorganisms dwelling within, can pave the way for chronic illnesses to take hold. Research continues to uncover how our resident bacteria, viruses, protozoa, and fungi impact physiological processes beyond digestion.

Many studies link dysbiosis — changes to the composition and function of these microbial communities from their normal state — to diabetes, autoimmunity, mental health conditions, and more. Scientists have discovered that many prevalent and pernicious diseases relate to a microbial imbalance in terms of diversity, disrupted communication pathways in the body, and overwhelmed immune function. For example, type 1 and type 2 diabetes involve immunoregulatory deficits, inflammation, and even autoimmunity — all of which relate closely to the gut microbiome's status. Depression and anxiety associated with exaggerated immune responses are also seemingly spurred by intestinal dysbiosis.

The microbiome interacts with and helps regulate the endocrine, immune, and nervous systems in exceedingly intricate ways. Microbes match both externally-produced compounds and endogenous ones, having profound implications for physiological harmony. When particular species deplete or dominate, lose diversity, and change profile compared to a healthy blueprint, the fallout can systematically stir inflammation, hamper key messengers, and disturb vital processes manifesting in chronic disease. Many scientists connect microbiome disruption to the puzzling rise of autoimmune diseases — conditions like Crohn's, rheumatoid arthritis, lupus, type 1 diabetes, multiple sclerosis, and more, where the body essentially attacks its tissues. Autoimmune disorders now affect 1 in 13 Americans and are growing, with rates doubling in certain diseases just since the early 2000s.

Research links gut dysbiosis to inappropriate immune training and faulty "self-recognition". Intestinal inflammation and permeability, both associated with microbiome imbalance, further trigger immune responses against body organs and tissues over time. Recurrent antibiotic use and dietary patterns depleting gut bacteria associated with immune regulation allow more aggressive, inflammatory microbes to occupy their niche. Viruses like Epstein-Barr take advantage, too. A child's chances of developing type 1 diabetes doubles, for example, with just one antibiotic course in the first 6 months of life.

Dysfunctional microbiome-immune communication, beginning early in life, sparks confusion over what's "self" versus foreign, instigating gradual auto-aggression. But microbiome balance helps maintain self-tolerance and tamp down overzealous autoreactivity through regulatory T-cells and blocking antibodies — potentially preventing attacks from triggering in the first place. When microbiome balance goes awry, the consequences can manifest as a full-blown chronic disease.

Many of today's most prevalent conditions relate intimately to enteric flora shifts that disturb homeostasis, perpetuate inflammation, confuse immunity, and foster multi-system dysfunction. Research continues unlocking microbiome influence across diverse chronic illnesses. As one salient example, type 2 diabetes incidence closely parallels rising obesity rates over recent decades. While poor diet and lifestyle habits are huge factors as well, scientists now recognize how certain microbes impact metabolic regulation and extraction.

More calories and carbs from foods promote weight gain versus others that enhance insulin sensitivity and fat metabolism, protecting against diabetes. Re-populating missing strains – our internal allies - makes sense. Likewise, Parkinson's disease shows gastrointestinal symptoms and inflammation decades before classic brain symptoms like tremors even arise. Tracing back the roots reveals pesticides and antibiotic use eradicating select neuronal-protective, dopamine-regulating bacteria, enabling alpha-synuclein proteins to misfold and nervous tissue to degenerate.

Leaky gut disorders and food sensitivities also tend to predate full inflammatory diagnoses like Crohn's or arthritis. Stress, infections, and Nonsteroidal anti-inflammatory drugs (NSAIDs) first bring intestinal permeability, allowing undigested proteins and toxins into circulation, prompting immune hypersensitivity against gluten, eggs, dairy, and other trigger foods. Allergies and joint pain later strike. Sealing tight junctions through diet or prebiotics potentially prevents dire autoimmunity. While the microbiome alone cannot singly explain complex, multifaceted conditions, it undeniably regulates processes underpinning chronic conditions of nearly all kinds affecting huge swathes of society. Restoring ecosystem balance and function with probiotics, prebiotics, and health-promoting lifestyle factors may ameliorate the foundations of disease and tip physiology back into homeostasis. Here, we explore some of today's major chronic illnesses for which microbiome science brings a deeper understanding of origins and opportunities for therapy via nutritional supplementation, FMT (fecal microbiome transfer), and more to correct what's gone wrong inside.

Topping the list as the number one killer of Americans is heart disease, which underlies heart attacks, arrhythmias, and strokes. Surprisingly, certain oral and gut microbes are now linked to circulatory disease processes enabling atherosclerosis or hardening/clogging of arteries feeding the heart.

Main culprits like Porphyromonasgingivalis and Tannerella species trigger systemic inflammation plus autoimmune reactions against vascular linings where cholesterol plaques stick and accumulate. Studies showcase greater arterial plaque deposition in those with periodontal pathogen-ridden gums and oral microbiomes promoting platelet activation, fibrin production, and clumping cascades, which can ultimately cut off blood flow through tissue oxygen starvation. Supporting healthy oral and gut flora balance promotes cardiovascular wellness. Regular exercise, a reduction in the amount of red meat that we eat, not smoking, and keeping our alcohol intake low all go a long way toward nurturing our cardiovascular health.

As mentioned, obesity, insulin resistance, and type 2 diabetes intertwine inextricably with microbiome disruption, which both promote and result from these epidemic metabolic conditions now topping global disease burdens. Select microbes mediate appetite signaling, glucose tolerance, lipid metabolism, oxidative stress, and more - tipping the balance towards or away from fat storage and are diabetogenic based on their activities and messages. When dysbiosis occurs, we lose "bug buddies" that help regulate weight-influencing hormones like leptin and ghrelin, which control hunger and satiety cycles.

We also harbor more craving-inducing fungi and bacteria that thrive on high-sugar, high-fat "Western" diets, which further the vicious cycles of obesity and insulin resistance known to underlie type 2 diabetes. Those McDonald's burgers and sugary Starbucks coffees are key areas to cut out problematic sugars and fats. Another scientific revelation involves the ability for select microbes to either promote or suppress oncogenesis – the formation of cancerous cellular abnormalities leading to tumor development.

For example, certain oral microbiome inhabitants like Fusobacterium stimulate enzymes called beta glucuronidases, which can reactivate cancer-causing toxins, potentiating risks while consuming free radicals and endotoxins helping prevent genetic mutations. The oral pathogen Porphyromonasgingivalis again appears to enhance cell proliferation and migration in pancreatic cancer while Fusobacterium nucleatum binds tumor growth factor beta, furthering disease. Gut bacteria produce pro- and anti-carcinogenic messaging molecules short-chain fatty acids (SCFAs) mediating gene transcription and immunity related to cancer development.

Harnessing bugs as terrain managers provide opportunities for oncological prevention and care. However, it's crucial to mention that there is no clear cause of cancer from this. Globally, scientists have been working for decades to understand the cause of this horrible disease. Still, in its early stages, the microbiome field holds promise for gauging risks, enabling detection, and modifying disease. Colorectal cancer (CRC) correlates strongly with the enrichment and prevalence of certain bacterial species, which generate genotoxins, make cells hyper-proliferate, and silence host immune responses against malignant growth. In those with colitis long term, CRC chances surge exponentially. Testing microbial inhabitants serves as a biosensor for trouble potentially brewing, offering chances to intervene through lifestyle habits and functional foods. Future applications and therapeutics related to microbiomics in oncology abound, from phage therapies to transgenic bacteria programmed to infiltrate hypoxic tumor cores. Continued research will clarify optimal strains to consume debris and decrease local inflammation while liberating immunostimulatory molecules to strengthen anti-tumor responses for adjunctive cancer care.

But let's explore more about neurodegenerative diseases. For example, in Parkinson's disease, aggregation of the alpha-synuclein protein forms characteristic clumps termed Lewy bodies, depleting dopamine-producing nerves profoundly disrupting motor control, cognition, and mood regulation. It turns out that certain bacteria possess the metabolic machinery to regulate alpha-synuclein expression along gut epithelium and to directly break down the amyloid aggregates themselves through bacterial amyloids and proteases – essentially delaying disease progression. Managing microbiome residents factors into neurodegenerative trajectories. Likewise, the amyloid beta accumulation characterizing Alzheimer's pathology relates to microglial activation patterns steered by intestinal microbiota through gut barriers, lymph drainage, and blood-brain exchanges.

Poor microbial richness impairs amyloid clearance mechanisms mediated partly via short-chain fatty acid byproducts and secretory IgA (these protect us from harmful pathogens and toxins), further enabling vascular damage and neurotoxic protein buildup. Improving diversity and anti-inflammatory species provides benefits. Prebiotic fibers serve as microbiome fertilizer, encouraging butyrate-liberating bacteria, which then assist the microglial clearance of neuron-clogging amyloid and manufacture key raw materials needed for healthy brain growth through synaptogenesis and myelination. Microbes feed brain cells and preserve function through such pathways, illustrating the promise of biotherapies in arresting chronic neurodegeneration by nourishing our brain's microbial allies. Intertwining tightly with neurological health, mental well-being also finds intimate connections to the assortment of bacteria populating our microbiome, especially within the enteric nervous system residing in the gut wall, considered by some scientists as a "second brain."

Gut microbes make many feel-good neurotransmitters like serotonin and dopamine, which are relied upon for stable moods, appropriate stress responses, and overall emotional balance. They also generate short-chain fatty acids, further influencing gene expression and impacting cognition. When dysbiosis occurs through factors like antibiotics or poor diet, we lose crucial psychobiotic microbes and their soothing molecular messengers. Opportunist strains more often thrive on stress hormones and sugar abundance, further promoting inflammation and exacerbating anxiety or depression-prone neural circuitry by disturbing key feedback loops between the gut and brain. This again highlights the importance of a balanced diet with plenty of probiotics and fiber. In recent years, more and more studies have been done on the natural product CBD.

CBD is known as Cannabidiol, which was first discovered in 1940. The cannabis compound CBD influences signaling along the vagus nerve, which connects the brain to the intestines. Through action upon CB1 and CB2 receptors expressed on immune cells, throughout the enteric nervous system and even on epithelial cells, CBD appears to exert broad anti-inflammatory, analgesic, and anxiolytic activity mediated in part by re-establishing homeostatic intestinal microbiota and gut wall permeability central to whole body and mind wellness. In one study, researchers gave probiotic supplements to patients with chronic fatigue syndrome, which commonly coincides with anxiety. Over the two months, almost three-quarters of subjects reported significant improvements in anxiety levels, which coincided with notable shifts in overgrowths of D-lactic acid-producing bacteria frequently aggravating brain fogginess and emotional distress.

Results lend further credence to psychiatry's increasing consideration of microbiome patterns both resulting from and playing into mood disorders – revealing bidirectional involvement of gut flora inhabitants as influencers and indicators of mental health status. It also highlights the need for more research on this fascinating natural extract, and we can use it going forward to help with more health issues, such as chronic pain.

Chapter 6

Diet, Lifestyle and the Microbiome

As we slightly explored in previous chapters, the foods we eat profoundly affect the trillions of microbes residing within our intestines. An unhealthy diet high in processed foods and lacking in fiber can decimate populations of beneficial bacteria while enabling disease-causing microbes to flourish. However, the right foods can shift the microbiome back towards balance and diversity, enhancing everything from digestion and weight management to mood and immunity – and the good news is that it's never too late to start. Let's explore how our dietary choices shape the intricate microbial ecosystems inside us.

Our gastrointestinal tract provides a range of ecological niches perfectly suited for colonization by complex communities of bacteria, viruses, and fungi. The human gut microbiome contains over 1000 diverse bacterial species, with densities reaching 1011 cells per gram of intestinal content. As our primary source of nourishment and energy, food residues passing through provide sustenance for hungry microbes. Different species have adapted to thrive on various components that escape digestion by our enzymes, from soluble fibers and resistant starches to proteins and fats. Thus, our long-term eating patterns directly impact microbial numbers and proportions. Research shows that high-fat, low-fiber Western diets decrease counts of beneficial Bifidobacteria and Lactobacilli compared to traditional high-fiber, unprocessed meals. Processed carbohydrates and added sugars also enable unrestrained growth of opportunistic bacteria like Clostridium difficile. This can trigger inflammation, change appetite signaling, and increase the risk for painful and annoying conditions like irritable bowel syndrome. Multiple studies demonstrate that dietary interventions rapidly shift microbial profiles – changes detectable within days. Simply reducing intake of animal fats and refined grains alters over 70% of gut species. Swapping processed foods for whole plant foods restores balance while suppressing pathogens.

Our everyday food choices hold sway over this complex internal ecosystem. Fiber-rich whole plant foods feed our microbiome the way nature intended - with prebiotics. In fruits, vegetables, nuts, seeds and whole grains, dietary fiber escapes digestion in the small intestine, reaching the colon intact, where specialized bacteria feast upon it. Insoluble fibers, like those in wheat bran, mainly assist with stool formation and bowel motility. Soluble fibers get fermented into short-chain fatty acids, including butyrate, propionate, and acetate – compounds that supply energy to intestinal cells, improve gut barrier integrity, enhance detoxification, decrease inflammation, and protect against colon cancer. This is why many people get their appendix removed. The appendix evolved over millennia to no longer be needed by modern humans because we no longer eat roots, grass or other roughage.

Furthermore, specific prebiotic foods exert unique effects based on their distinct fiber profiles. For example, commensal Bifidobacteria thrive on fructooligosaccharides in onions, garlic, asparagus, and other vegetables, allowing them to produce nutrients for enterocytes and crowd out pathogens. Bananas contain inulin-type fructans used by Faecalibacterium prausnitzii to generate butyrate and support colonic epithelium health. Tomato fibers enhance edema-fighting Roseburia intestinalis and anti-inflammatory strains. This enormous fiber variety in whole plant foods enables us to nourish diverse commensals based on our individual needs selectively. Another key mechanism for plant foods involves polyphenols - compounds that get broken down by certain gut bacteria into bioactive, health-promoting substances.

Phenolic compounds in cocoa, green tea, coffee, berries, nuts, and extra virgin olive oil seem particularly important for enhancing microbial richness. Polyphenol-metabolizing bacteria even hold promise for preventing dysbiosis and related disorders, according to research. This interaction between plant phytochemicals and intestinal microbes only recently came under scientific scrutiny – more research is always being done in this area.. Specific diets leveraging whole plant foods and polyphenols have proven valuable clinically for correcting dysbiosis and restoring eubiosis (opposite of dysbiosis). Plant-based Mediterranean plans emphasize fruits, veggies, herbs, nuts, legumes, fish, and olive oil while limiting red meats and refined carbs. Studies verify this diet increases microbial diversity and beneficial taxa compared to higher-fat Western meals. The traditional Okinawan diet centered on vegetables, sweet potatoes, green tea, and soy foods has been linked to longevity and ideal gut flora proportions.

Modern diets like the Pegan Diet combine elements from Paleo and Vegan approaches to support optimal microbiome balance. Appropriate plant-focused diets provide ecology-appropriate sustenance to our micro-friends. Beyond dietary patterns, our meal timing can also dramatically impact rhythms in the gut microbiota through crosstalk between intestinal bacteria and our circadian clock. Microbes display oscillations in activity and functionality over 24-hour cycles, patterns regulated by our feeding/fasting sequence. Time-restricted eating approaches and intermittent fasting regimens improve microbial diversity and health markers by allowing extended rest periods without incoming nutrients.

This fasting-induced “ebb and flow” seems critical to overall microbiome fitness, just as sleep offers our brain cells time for repair and recovery at night. Of course, sourcing high-quality ingredients deserves mention, given rising concerns over antimicrobial residues, heavy metal contamination risk, and depleted nutrient density in our modern food chain. Organic produce minimizes pesticide exposure, supporting microbial integrity (and avoids GMOs with unknown ecosystem impacts).

Green-fed beef provides more conjugated linoleic acid to dampen inflammation than conventional options. Wild fish offer superior omega-3 levels, benefitting enterocyte function and barrier permeability over farmed varieties. However, it’s important to acknowledge that many people don’t have access to or the financial means to follow this kind of diet. While personalized to one’s needs and access, consciously curating our food sources promotes microbial symbionts through quality components while limiting hazardous factors.

Our gastrointestinal tract co-evolved in partnership with microbiota over millions of years into a mutually flourishing ecosystem encapsulated within. Eating patterns mimicking our ancestral hunter-gatherer diets - whole food, fiber & phytonutrient-rich, organic, unprocessed - serve to nourish commensal symbionts that supply essential compounds and crowd out pathogens. Dietary choices impact everything from satiety signals and mental outlook to glycaemic control and cancer risk - influence largely mediated through the gut microbiome. Hippocrates proclaimed, “let food be thy medicine” - The wisdom we now understand extends to nourishing the multitudinous micro-life within.

As we explore lifestyle factors shaping the gut microbiome beyond diet, physical activity is a powerful modulator of microbial populations and inflammatory status. Athletic individuals possess a wider diversity of gut bacteria compared to sedentary peers. Exercise exerts multi-faceted prebiotic, metabolic, and immunologic effects, enhancing overall microbiome fitness.

Yet chronic inactivity poses an enormous threat in our modern, screen-dominated society. Over 60% of American adults report no weekly exercise, but even modest activity helps avoid dysbiosis based on swathes of reliable research. Such Studies demonstrate just 30 minutes of moderate jogging increases counts of beneficial Lactobacillus, Bifidobacterium and Roseburia taxa. These genera improve IBS symptoms, enhance barrier function, raise T-regulatory cells, and suppress inflammation. Aerobic activities like running also limit blood flow to digestive organs, triggering exercise-induced ischemia that alters oxygenation, nutrient availability, and pH locally.

This provides a competitive advantage to anaerobic microbes over aerotolerant groups less equipped for such conditions. Ultimately, microbial "interval training" occurs through alternating ischemia and reperfusion during physical movement. Weight-bearing exercise further impacts the gut lining directly via muscle-enterocyte crosstalk. Skeletal muscle contractions release myokines systemically – substances that enhance intestinal cell proliferation, mucus secretion, and tight junction protein expression. Resultant improvements in barrier integrity assist symbionts while restricting pathogens and antigens. Myokines also regulate endocannabinoid tone, appetite signals, and fat metabolism. Overall, the profound benefits of routine movement for our microbiome (and health generally) cannot be overstated. Another lifestyle factor involves natural environments, outdoor recreation, and the human skin microbiome.

It seems our modern society has underappreciated the benefits of spending time immersed within diverse external ecosystems – exposures evolutionarily fundamental for Homo sapiens spanning 200,000+ years. Yet urban living severs this contact. Americans average a mere 7 hours per month outdoors in nature while devoting over 50 hours monthly to screens.

What impact does this acute deprivation have on our collective microbiomes and health? Researchers now actively investigate this question with intriguing early findings, such as decreased memory ability and inability to focus. Comparative analyses find individuals who regularly spend time in forests and green spaces harbor more diverse skin commensals versus urbanites. Rural environments seed a wealth of new microbes onto our epithelial surfaces – exposure called transient microbial inoculations by scientists. Frequent inoculations assist the cutaneous immune system in maintaining immunological tolerance to avoid hypersensitivity and inflammation against benign microorganisms omnipresent around us.

Those deprived of such inputs instead show increased rates of allergies, atopic dermatitis, and even metabolic disorders over recent decades as cities expand. Our innate microbial ecosystems require periodic replenishment from surroundings ideally. Specific activities like hiking or swimming provide direct skin contact with soil, foliage, and water - wealthy sources of unfamiliar microbes. Just 20 minutes immersed at a beach plants nearly 100,000 bacteria onto the hands and body, according to research. Subsequent microbial mixing and gene transfer likely assist in enriching protective skin commensals and educating our local immunity. Studies even show such transient flora get transported by APCs to distal lymph nodes, introducing themselves to the greater immune system. Similar dynamics presumably occur along the GI tract with oral intake and inhalation of ubiquitous micro-life in our midst. We have only begun elucidating such processes, but findings reinforce lifestyle choices, bringing us into diverse natural settings as microbiome-supportive. Beyond the outdoors, our household environment also greatly influences daily microbial exposures.

Floors, faucets, countertops, and other surfaces harbor millions of bacteria - populations mixing through shedding from inhabitants plus tracked-in soil and environmental microbes. Beneficial species diffuse indoors on dust particles and through open windows. Traditional homes built from natural materials like wood, stone, and adobe bricks support more ecologically balanced ecosystems internally versus hermetically sealed modern dwellings full of plastics, paints, and industrial cleaners. The effects of these environmental differences on human microbiomes prove quite profound clinically.

However, we mustn't live in overly clean, sterile environments. Curiously, multiple studies demonstrate living on traditional farms correlates with lowered risk for allergic disorders, asthma, and inflammatory conditions. Frequent contact with livestock, hay, soil, and fermented foods enriches the biodiversity of rural microbiomes – effects detectable in comparison to urban households. Specific bacterial taxa overrepresented on farms assist in immune education and preventing hypersensitivities. Clean drinking water also shapes gut flora, preventing dysbiosis.

Our surroundings contain a natural background microbiome subtly influencing our resident populations. Yet modern society's severe detachment from ancestral environments has not come without consequence. For example, obsessive focus on sterilizing our living spaces destroys beneficial environmental microbes once omnipresent indoors and out. Antibacterial cleaning agents and harsh detergents suppress commensals while encouraging adaptive pathogens through repeated use. And by spending over 90% of our lives sealed inside artificial structures, often with recycled air, we lack adequate exposure to the natural microbial world.

Researchers now actively investigate corrective measures through architectural modifications allowing green space integration, microbial exchange, and biodiverse landscaping in urban planning. Our health depends deeply on balanced contact with microbial life surrounding us. Even our connections with pets and other animals deserve mention regarding bidirectional microbiome enrichment. Ownership of dogs dramatically diversifies the skin microbiome through shedding, cuddling, and shared living spaces.

Dog owners also spend more time outdoors exercising and socializing – lifestyle factors aiding balanced flora as we have covered. Beyond dogs, animal-assisted therapy utilizing horses, dolphins, and farm animals also enriches protective microbes through similar social intimacy, emotion regulation, oxytocin release plus outdoor exposures. Soldiers suffering from PTSD show gut dysbiosis correctable through equine therapy pairing microbial seeding during grooming and riding with limbic system modulation, for example.

In essence, complex microbial communities cooperate functionally as “mega-organisms" in which individual members relay signals coordinating collective behaviors from nutrient harvesting to protection against pathogens and opportunists. Our social world as humans seem to operate through similar group dynamics and alliances – making these bidirectional communication channels with microbes quite harmonious and deeply rooted. Social connections serve as conduits supporting the resilience of microbial networks inside us.

Shifting gears to psychological lifestyle factors, a wealth of compelling research illuminates the influence of stress, mindset, and emotions on the gut microbiome. Beginning with stress, hypothalamic activation of the sympathoadrenal and HPA axis initiates a conserved physiological cascade marked by elevated heart rate, blood pressure, respiration, vigilance, and stress hormones, including cortisol and adrenaline.

This fight-or-flight mobilization diverges energy resources away from anabolic functions like digestion and tissue growth towards crisis response temporarily until threats are resolved. Only then can homeostatic parasympathetic activities resume. In modern society, chronic elevation of glucocorticoids, catecholamines, and inflammatory signaling emerges with recurring psychosocial stressors like job strain, financial woes, relationship conflicts, or perceived time pressures. Such sustained stress physiology decimates beneficial microbial populations unable to handle constant hormones and oxygen radicals.

Protective groups like Bifidobacteria and Lactobacilli decline while the influx of carbohydrates and fats (due to cravings and overeating) allows the expansion of pro-inflammatory Proteobacteria for example. A leaky gut also develops as tight junctions unravel and mucus production lags. Eventually, this lingering breach can trigger wide-ranging problems. Research confirms stressed individuals demonstrate gut permeability, microbial imbalance, and depleted GABA-producing Bifidobacterium long before any gastrointestinal symptoms materialize. Minimizing lifestyle stressors clearly serves as imperative for prevention. Targeting HPA overactivity directly through adaptogenic herbs, biofeedback, yoga, nature immersion, and counseling represent additional avenues to curb cortisol fallout in the gut. But perhaps most fascinating are recent discoveries surrounding probiotic strains with inherent anti-stress capabilities being demonstrated in trials now. Dovetailing with stress, our cognitive mindset, self-perceptions, beliefs, and emotional patterns influence microbial communities.

For example, several studies demonstrate depression and anxiety correlate with decreased gut diversity, altered immune signaling, and greater cortisol awakening response compared to non-depressed controls. Microbes modulate neurotransmitters, immune messengers, and enteric nerves that influence mood regulation bi-directionally. While speculative presently, thought patterns and emotional experiences could further signal directly to microbial participants. Initial evidence in humans and animals suggests destructive emotions like anger, hostility, and resentment relate to increased inflammatory T-cells and cytokines systematically along with depletion of beneficial bacteria. Conversely, positive emotional states and self-compassion may support homeostasis and symbiont strength through reciprocal nervous system pathways and endocrine connections under investigation. Practicing self-love could assist our mutualistic microbes and make conscious efforts not to overwork in our jobs.

With the rise of TikTok, Google, and other internet platforms, people are taking it upon themselves to diagnose themselves or others with certain conditions or mental or physical issues. Emerging study of the "placebo effect" and associated conditioning, expectations, and predictive beliefs reveals the potential to override physiology stemming from conscious and subconscious processing. Given the extensive capability of microbes to manipulate brain activity and behavioral responses already established in animals, scientists now assess whether probiotics could mediate placebo benefits through “belief-assisted supplementation”. Since roughly half the effect of certain medicines derives from patient expectations pre-treatment, leveraging such conditioning through repetitive thought could boost outcomes. Perhaps one day, prescribing pure sugar pills alongside probiotics may enhance results through additive psychosocial and psychobiotic synergies!

In summary, conventional lifestyle dogma focuses heavily on diet and exercise for health promotion - measures certainly invaluable for supporting balanced microbiome populations. However, we must widen frameworks to acknowledge the equally substantial impact of physical activities in natural environments, social integration, animal bonds, chronic stress avoidance, emotional well-being, and psychological resilience demonstrated in studies. Sampling microbial diversity found in ancestral habitats serves our commensals well through inoculations. Stress and isolation disrupt community structure, while compassion and connection reinforce it.

The conscious mind likely exhibits untapped potential to communicate downwards by signaling our microbes intentionally. While research remains preliminary in places, lifestyles allowing our innate microbial selves to flourish symbiotically clearly deserve integration for optimal well-being.

Chapter 7

Prebiotics, Probiotics and Synbiotics

By now, you have a good understanding of the complex microbial communities residing within your gut, skin, mouth, and other bodily areas. These microbes wield tremendous influence over many aspects of your health. From digestion and metabolism to immunity and even mood, our microbiomes play pivotal roles. We've mentioned a few times now that a balanced microbiome is a remarkably diverse, resilient ecosystem.

But when thrown out of balance by things like antibiotics, stress, or an unhealthy diet, trouble brews. Beneficial microbes dwindle while more harmful varieties flourish. This dysbiosis lies at the root of most chronic illnesses today. Fortunately, resetting equilibrium in your inner ecology is possible through the strategic use of prebiotics, probiotics, and synbiotics. Synbiotics are mixtures of probiotics (helpful gut bacteria) and prebiotics (non-digestible fibers that help these bacteria grow).

If all of this terminology is confusing, don't worry! It's my goal to ensure that this terminology is accessible to people who don't have a detailed knowledge of science or biology. Prebiotics are specialized plant fibers that humans can not digest. But certain bacteria in your microbiome can! Prebiotics selectively feed beneficial bacteria, helping them flourish. The word itself tells the story - "pre" for preceding and "biotic" for life. Simply put, prebiotics set the stage for healthy microbiome function. They stimulate the growth and activity of bacteria that offer advantages to the host. Bacteria use prebiotics for fuel to multiply and make vitamins, immune signals, and other beneficial compounds. Common prebiotics include inulin, fructooligosaccharides (FOS), galactooligosaccharides (GOS) and resistant starches. Foods rich in prebiotic fibers include garlic, onions, leeks, asparagus, bananas, whole grains, and legumes. In 1921, Japanese scientist Minoru Shirota discovered unusual bacteria that thrived on prebiotics in cabbage.

He went on to launch a commercial probiotic drink using these cultures. Today, scientists recognize that such fibers selectively promote protective bacteria like Lactobacilli and Bifidobacteria strains. Meanwhile, less favorable bacteria lack the right enzymes to metabolize most prebiotics. Essentially, prebiotics nourishes good bacteria to keep bad bacteria in check through competitive exclusion. Studies show prebiotic consumption boosts immune function and gut barrier integrity while lowering inflammation.

In one trial, pregnant women took GOS prebiotics for six weeks before labor. Their vaginal microbiome contained more good Lactobacilli while pathogenic bacteria decreased. This led to a significantly lower risk for premature rupture of membranes, infections and other birth complications. Multiple studies confirm that prebiotic intake reduces diarrhea from gastrointestinal infections. Athletes who use prebiotics experience less upper respiratory infections and reduced allergy symptoms.

One key point is that you must continually consume prebiotics to maintain benefits. While probiotics can permanently colonize your gut, prebiotics are non-living fibers easily flushed from the system. Stop eating them, and bacterial levels will quickly fall back to baseline. Much remains unknown regarding ideal daily prebiotic intakes across populations. Most researchers recommend aiming for 5-10g of mixed prebiotics from whole foods. Unfortunately, a whopping 96% of Americans fall far below those levels.

Now that you understand how prebiotics selectively feed good gut bugs, let's explore probiotics more closely. These are actual live strains of beneficial bacteria that colonize your microbiome. The World Health Organization (WHO) defines probiotics as "live microorganisms which, when administered in adequate amounts, confer a health benefit on the host". But keep in mind that not all bacteria deserve superhero status! Human research identifies a select handful of genera that provide consistent advantages, including Lactobacillus, Bifidobacterium and Saccharomyces.

There are also heroic species within less familiar genera like Streptococcus and Bacillus. Probiotic brands name each unique strain by its genus, species and alphanumeric designation. For example, Lactobacillus rhamnosus GG (often abbreviated LGG) is one of the most clinically validated and effective probiotic strains ever identified. Meanwhile, certain strains of E.coli and Klebsiella can actually be pathogenic. So, the species designation matters greatly! Through years of research, pinpointing which strains carry meaningful benefits has proven complex. It seems that probiotics act in a highly selective and strain-specific manner. Those showing promise in the lab may not replicate results in human users. Within a species, individual strains use diverse mechanisms in different host environments.

The dosing regimen and delivery matrix also determine how effectively strains colonize our inner ecosystem. With over 5000 studies examining probiotics, ferreting out useful conclusions requires examining the collective evidence. Meta-analyses help assess bonafide advantages linked to specific strains. Such reviews consistently associate selected probiotics with enhanced immunity, digestion, nutrient absorption, weight control, and metabolic health. They also help restore normal gut permeability and microbiome equilibrium. Probiotics temper inflammation while reinforcing gut barrier strength to keep pathogens at bay. But we have yet to characterize the core mechanisms driving positive change. Future advances in gene sequencing will uncover more about how heroic strains impact genetic expression of human cells. Ongoing research also looks at how certain species coexist and signal each other. This may allow the creation of custom probiotic cocktails that synergize within a community.

One nagging question remains – are benefits limited to the supplement itself? Or do friendly flora take up long-term residence within users? Early evidence suggests that select strains can permanently colonize the gut. But most seem to transiently pass through while they exert positive influences. Despite dwell time, probiotics remain powerful tools for enhancing and maintaining a balanced microbiome ecosystem. For magnifying microbiome benefits, combining prebiotics and probiotics as synbiotics packs a real punch. Synbiotics are mixtures of complementary prebiotics and probiotics designed to boost beneficial flora. These dynamic duos work together to improve the colonization, survival, and activity of supplemental bacteria. Specifically, the prebiotic fibers selectively nourish the paired strains of probiotics.

This preferential feeding helps probiotic cells replicate faster to achieve greater numbers with enhanced durability. Prebiotics support larger probiotic populations that diversify the gut community by providing nurturing growth substrates. Enhanced diversity contributes stability while crowding out unfavorable species. Research substantiates that mixing prebiotics and probiotics as synbiotics increases positive outcomes. Synbiotics increase good Lactobacilli, and Bifidobacteria counts better than either alone. In pregnant women, synbiotics slash UTI rates, lessen pain, and accelerate recovery after C-sections. Athletes using synbiotics withstand intensive training better while avoiding infection. Combining prebiotics and probiotics provides superior symptom relief for GI disorders like irritable bowel syndrome (IBS) and infectious diarrhea. One clever study created a novel synbiotic using resistant starch and Bifidobacterium. Rats eating this combo had more adaptations to colon cells that transport blood sugar than controls.

The synbiotic group also showed lower blood sugar, insulin, and anti-inflammatory T cell changes. This hints at microbiome-based strategies for managing diabetes and metabolic disease! Before getting carried away, a reality check is necessary. Synbiotics are no quick fix-all. Products vary enormously in strains, dosage, price, viability, and format. Blindly mixing common probiotic brands with random prebiotic fibers fails to utilize strain-matched synergy. Also, more diverse strains may not integrate as smoothly as pairing a single species with its preferred substrate.

Rational synbiotic design must match prebiotics to the exact species' metabolic capacities. This requires detailed genomic analysis mapping each probiotic strain's carbohydrate enzymes and transporters. Companies also need stricter quality controls to ensure product viability through complex delivery matrices. But when specific prebiotic-probiotic pairings harmonize effectively, microbiome restoration may progress further. Harnessing complementary synbiotic suites offers ecological approaches for preventing dysbiosis-related disorders. Perhaps in the future, clinical isolation will identify signature strains from each person as custom probiotics. Personalized prebiotics would then selectively propagate your heroic flora for augmented ecosystem stability. As we delve into the validated advantages linked to prebiotics, probiotics, and synbiotics, consider that clinical research typically focuses on alleviating disease. But microbiome balancing provides far more than symptomatic relief for those already unwell. Strategically harnessing good bacteria grants protective, performance, and longevity benefits starting in our youngest years.

When pregnant mothers take certain probiotics and prebiotics, their newborns have significantly lower rates of allergies, eczema, and respiratory illness through age 5. Athletes using prebiotics perform better with heightened endurance, strength, and faster race times while avoiding upper respiratory infections. Optimizing gut flora helps us achieve healthier body composition by preferentially burning fat over carbs. One study found that just 30 days of prebiotic intake boosted good gut bugs by 130 times and doubled microbial diversity. This led participants to lose 4 additional pounds of fat without any other changes.

We also know that good bacteria produce key vitamins, activate nutrients, and break down toxins. They protect the gut barrier that keeps pathogens away from the bloodstream. And evidence proves repairing leaky gut with probiotics can resolve autoimmune disorders. One incredible investigation showed that transferring gut bacteria from athletic mice into inactive mice led to greater endurance capacity after just 6 weeks! Even neurochemicals involved in mood regulation trace back to bacterial actions far downstream. It seems microbes influence far more than just digestion and immunity! While consumer marketing focuses heavily on digestive issues, clinical applications for prebiotics and probiotics run remarkably deep. They help prevent systemic inflammation underlying nearly all chronic diseases. Certain strains slash allergy and asthma symptoms while decreasing the risk of respiratory infections. Both kids and adults using probiotics average nearly 20% fewer colds and flu annually. With consistent use for 3 months, respiratory disease severity and duration plummet even more. Those prone to sinus troubles, bronchitis or pneumonia may benefit greatly.

For cancer prevention, probiotics balance estrogen metabolism to suppress tumor growth. A bout of antibiotics decimates good flora and doubles your odds for colon cancer. But those who consume yogurt kefir or take probiotics run a 25-42% lower risk. Compounds secreted by certain species even hamper angiogenesis, potentially starving developing cancers by blocking their blood supply. When tumors spread aggressively, probiotics enhance efficacy and temper the side effects of chemo drugs.

Radiation therapy recipients also suffer far fewer adverse events taking select probiotics. Because gut imbalance allows undigested proteins and pathogens to leak into the bloodstream, probiotics help prevent type 1 diabetes. Evidence proves they also lower inflammatory factors underpinning type 2 diabetes and metabolic syndrome. Probiotics boost insulin sensitivity by altering gut permeability and microbiome-derived metabolites. For diabetics already taking metformin, one particular strain improved drug efficacy thrice. Markers of oxidative stress, glycated hemoglobin, and insulin resistance plunged substantially. Those eating probiotic yogurt had 25% less likelihood of developing diabetes over a decade than peers.

But we've only scratched the surface of microbiome medicine for metabolic regulation! With obesity at epidemic proportions, researchers eagerly explore weight loss apps. Specific strains turn on fasting cues while suppressing appetite hormones to limit overeating. Obese adults shed 8-10% more weight by adding probiotics/prebiotics for 12 weeks than diet/exercise alone. Waistlines shrink further when used for over 6 months. Rodent data reveals certain butyrate-producing probiotics prevent and resolve fatty liver disease. By calming chronic inflammation driving obesity progression, microbial therapies offer adjuncts to behavioral modification.

For digestive disorders like IBS, IBD, and acute infectious diarrhea, probiotics deliver faster, more complete symptom relief. Abdominal pain, cramping, bloating, constipation, and loose stools all improve significantly. In one trial, a Bifidobacteria probiotic resolved diarrhea for 80% of those with IBS. It also eliminated bloating for 83% compared to 49% taking a placebo. The same species cut ulcerative colitis flares in half over one year. Prebiotics and synbiotics also repair leaky gut underlying digestive troubles. This helps alleviate food reactions, nutrient malabsorption, and systemic inflammation. With multidrug-resistant organisms threatening modern healthcare, probiotic strategies have gained interest. Certain Lactobacilli inhibit drug-resistant strains of Staph, E.coli, Klebsiella, Salmonella, Clostridium, and more. Specific strains disable virulence genes while stimulating innate defenses. Powerful Bifidobacteria produce acetate that kills the lethal hospital pathogen C. difficile. One synbiotic slashed gut colonization by this stubborn infection better than leading antibiotics. Such ecological methods may soon substitute our failing pharmacological model.

With far lower rates of autoimmune disorders, some posit that helminthic gut parasites mediate inflammation. Support comes from trials intentionally infecting people with whipworm eggs to treat conditions like IBD, MS, and autism. In developed nations, our ultra-hygienic lifestyles preclude exposure to such organisms. However, probiotics like the parasites may furnish immunomodulatory metabolites. Scientists continue investigating helminth/microbiome interplay in immune health. Research constantly uncovers new therapeutic applications for supplementing heroic flora. But remember – ingesting microbes alone fails to override poor diet and lifestyle choices, damaging microbiome resilience. Probiotics aid antibiotic recovery or crowd out acute infection. Prebiotics amplify good bacteria, keeping troublesome species suppressed through competitive inhibition.

But long-term symbiosis with our microbial minders requires feeding them properly while avoiding ecosystem disruption. Promoting diversity through high fiber intake and fermented foods sustains equilibrium most naturally. Our ancient ancestors likely hosted over 1500 species! So modern living sabotages community stability vital for wellbeing. Reinoculating with key commensals fine-tunes communication along the gut-brain network. But permanently seeding diversity resembling ancestral norms may necessitate broader inputs. Beyond isolating signature responder strains, we must characterize the contextual networks that nurtured their function over countless centuries of hominid evolution.

Chapter 8

Antibiotics: Damage and Dysbiosis

Modern medicine relies heavily on the use of antibiotics to treat bacterial infections. However, antibiotics are antimicrobial weapons designed to kill infection-causing microbes or inhibit their growth. Unfortunately, when we deploy antibiotics, they often cause extensive collateral damage to the complex microbial communities within us.

Let me explain it simply: antibiotics are like bombs in a city. When targeting criminals and enemies, they can eliminate threats, but they also destroy infrastructure, homes, and innocent lives. Similarly, when antibiotics attack infection-causing microbes, they also wipe out beneficial bacteria that are integral to the structure and function of our inner ecosystem. This antibiotic fallout can have devastating effects on the diversity and resilience of our microbiome that may last months or even years after a course of antibiotics. Alexander Fleming, the Scottish scientist who discovered penicillin in 1928, was the first to observe antibiotic resistance amongst bacteria and warn of overuse.

Yet modern medicine proceeded to mass produce and liberally prescribe antibiotics without much caution. We used them as cure-alls while neglecting their disruptive impact on delicate microbial balance within our bodies. By the late 20th century, antibiotic overuse had led to the evolution of superbugs - dangerous bacteria resistant to most known antibiotics. The drugs that were scientific marvels had been rendered increasingly ineffective against the microbial threats they aimed to conquer.

This is known in the medical and scientific community as “antimicrobial resistance”. It highlights the need for a more ecological, microbiome-centric approach to infection treatment. Changes in how humans can buy antibiotics is also influencing this. As pharmacists and other “non-doctors” are being permitted in many countries to prescribe antibiotics, there is increasingly a worry that it’s speeding up global antimicrobial resistance. One that weighs antibiotic therapy against long-term damage to beneficial bacteria is crucial for human health and disease prevention. But how do they work? Antibiotics indiscriminately kill susceptible bacteria through different mechanisms of action. Some inhibit bacterial cell wall construction, others interfere with protein synthesis or DNA replication. This causes extensive loss of microbial life from our body's ecosystems.

Beneficial symbionts that help regulate metabolism, train our immunity, produce vitamins, protect against pathogens, and more are wiped out. Resident bacterial communities get altered beyond recognition. Meanwhile, some dormant spores and inherently resistant microbes selectively persist - rapidly multiplying to exploit ecological voids post-antibiotics. This decreases diversity and allows potentially harmful pathogens to colonize areas like our skin, mouth, and intestines unchallenged. Disrupting community dynamics through antibiotic use is akin to clear-cutting an old-growth forest brimming with diverse flora and fauna. Multi-layered habitats built up over the years are demolished overnight. Destruction at this scale leaves soil unstable and barren - vulnerable to invasive weeds. Slow, natural regeneration becomes uncertain. Much like our gut lining post-antibiotics, stripped of its microbial barrier that defends against toxins and inflammation. Recovery depends greatly on which bacteria remain and environmental factors influencing ecosystem renewal. After antibiotics were withdrawn from our inner world, founder populations left behind the seed of very different communities.

Meanwhile, our depleted microbiome lacks resources for displaced beneficial bacteria to reestablish themselves. This magnifies dysfunction, causing further stress and inflammation. The extensive elimination of beneficial microbes by antibiotics inflicts immense damage upon our native microbiome. This condition, known as dysbiosis, leaves it severely imbalanced with reduced diversity. As integral modulators of our immune function, digestion, hormones, and more - this disruption of our microbial composition has profound effects on systemic health. It sabotages vital processes within us, increasing susceptibility to infections, allergies, obesity, and other inflammatory disorders.

When we undermine our microbiome, we undermine our overall disease resilience at a genetic level. Many beneficial symbionts contain specialized genes that help us digest fibrous plant matter, make vitamins, regulate weight, fight infections, and even combat anxiety. Losing these bacteria through antibiotics diminishes the availability of such microbial genes within our gut. Meanwhile, pathogen-promoting genes in species like C. difficile flourish - amplifying risk. This alters gastrointestinal gene expression towards a pro-inflammatory state that impairs our well-being. For instance, enzymes produced by certain Firmicutes bacteria help release anti-anxiety metabolites and antioxidants from plant foods that reduce stress. But often, 90% of a person's gut Firmicutes disappear post-antibiotics - causing chemical coping deficits. Similar enzyme-based microbial assistance enables us to fully utilize nutritious, fibrous carbohydrates in vegetables, lentils, and fruit. Excess carbs ferment into abdominal bloating without key starch and complex sugar digesters. Metabolic regulation falters as well. Microbes produce short-chain fatty acids that trigger intestinal cells to release peptide hormones that reduce blood sugar spikes after meals. Antibiotics hinder this glycemic control.

One of the gravest dangers of microbiome disruption is how it enables specific pathogens like Clostridium difficile to colonize our gut extensively. C. difficile forms highly resistant spores that withstand antibiotics while surrounding bacteria perish. With competing microbes eliminated, C. diff balloons from a contained niche to a dominant pathogen - releasing toxins that damage intestinal cell walls.

This causes inflammation, diarrhea and life-threatening conditions like pseudomembranous colitis. Cases of severe C. difficile infection have dramatically risen due to antibiotic overuse. From less than 50,000 a year in 1996 up to 500,000 cases by 2014. Meanwhile, protective commensals that normally fortify mucosal barriers are decimated by antibiotics. Symbionts like Faecalibacterium prausnitzii release anti-inflammatory compounds that nurture gut cell integrity to block leaks. But F. prausnitzii is extremely vulnerable to antibiotics - with numbers dropping 95% or more with therapy. This defense loss leaves our enterocytes exposed to toxins, antigens, and LPS endotoxins from new invasive pathogens

. It triggers allergic reactions and sparks fiery cascades through our intestinal wall's cytokine chemical alarm system - calling white blood cells to combat infection threats erroneously. Such severe inflammation post-antibiotics taxes our immunity itself. 70% of our immune system surrounds the intestines, constantly communicating with microbes to hone defenses. Key T regulatory cells that reduce intestinal reactivity depend on short-chain fatty acid byproducts of fiber digestion by symbionts. Losing these SCFAs weakens tolerance, causing immune reactions to foods and oral antigens. This increases the onset of eczema, asthma and seasonal allergies after antibiotics. Infants also face high risk since their developing immune system relies more extensively on gut microbes for education and regulation during initial development.

Disrupting this crucial bi-directional dialogue early in life handicaps the balanced immune programming needed to prevent hypersensitivities. Hence, antibiotic exposure while young increases susceptibility to many inflammatory and autoimmune disorders. Antibiotic-induced loss of microbiome diversity also dismantles intricate checks and balances on our metabolism and fat storage. It enables certain microbes like Erysipelothrix to dominate while key beneficial regulator species dwindle. Erysipelothrix has genes that hijack liver lipids into fat production. Unchecked, it causes more calories to convert into body fat rather than fuel or heat. This metabolic dysfunction promotes weight gain. One study found infants given antibiotics before age 6 months became 22% more prone to childhood obesity.

Antimicrobial compounds from symbionts usually protect against such problematic species as well. But with keystone bacteria removed, resistant opportunists exploit newfound niche space and alter host signaling. For example, Akkermansia muciniphila is an intestinal symbiont that maintains weight by enhancing mucus barriers and gut peptide release. But it relies on partner Bifidobacteria species for sufficient growth. If those allies perish from antibiotics, A. muciniphila also declines - unable to impart its protective effects. This allows obesogenic bacteria to ignite appetite and fat production. Ultimately, antibiotics cause microbial bankruptcy within our gut. Special assets provided by once-abundant investor species are lost. The niche infrastructure they sculpted internally erodes away, leaving few settlements for rebuilding. Beneficial symbionts hovering at the brink of detection limits cannot readily repopulate a dysbiotic gut environment. Growth stalling factors persist - like lack of metabolites key species require. Or, they cannot withstand inflammation due to lost cohorts that formerly buffered local reactions.

Full microbiome recovery requires gradually reintroducing missing keystone strains while promoting the growth of helper groups in sequence. Providing biochemical signalling, physical scaffolds, and grazing defenses that enable step-wise community restoration. Inoculating with keystone F. prausnitzii is ineffective since it relies on Bifidobacteria metabolites. So Bifido species must be fostered first to bring back F. prausnitzii durably. Through layered revitalization, gaps left by antibiotics slowly heal until diversity and function balance. But preventing such damage in the first place remains imperative. The wonders of microbial traditions passed down generations slam shut with just a few doses of antibiotics. Forcing them to open again proves extremely challenging and will become a globally-known issue over the decades to come, but we empower ourselves with this knowledge now.

Chapter 9

Supporting Microbiome Diversity and Resilience

Getting a clear picture of the current state of your microbiome is the critical first step on the journey toward balance and resilience. Just as you wouldn't set out on a road trip without checking the fuel gauge or tire pressure, taking stock of the trillions of microbes inhabiting your gut before making lifestyle changes or taking supplements is important. The microbiome testing industry has exploded in recent years as researchers reveal connections between the gut microbiome and nearly every aspect of human health. From inflammatory bowel disease to anxiety, obesity, and even Parkinson's disease – imbalances in the delicate microbial ecosystem within us are implicated. While consumer microbiome tests remain controversial and their clinical utility is debated, they may provide helpful directional information and motivation for those seeking to cultivate a flourishing internal garden. Sophisticated laboratory techniques like high-throughput genetic sequencing allow the identification of microbial fingerprints and the Mapping of microbial neighborhoods with precision. However, even simple at-home biomarker tests can cast light on the balance of good bugs and bad bugs.

Tracking health-promoting short-chain fatty acids, inflammation-provoking lipopolysaccharides, and other microbial metabolites over time provides insight into how diet, stress levels, sleep, and lifestyle changes nudge the microbiome toward, or away from steady-state equilibrium. The old adage "you can't manage what you don't measure" certainly applies to the microbiome. Getting baseline biomarker data serves as a benchmark to track progress and provides accountability for those embarking on intensive gut healing protocols. Of course, numbers never tell the whole story regarding complex microbiome health.

We must artfully balance the quantitative with more intuitive qualitative assessments. Tuning into signals from our sensing self allows us to judge the effectiveness of microbiome promotion efforts. Energy levels, bowel movements, bloating, and even cravings provide insight no laboratory test can match. A good starting point involves rating physical and emotional markers using simple symptom questionnaires before and after interventions like yoga, adopting intermittent fasting, or challenge testing to reveal individual food intolerances.

While the scientific language of diversity, gene richness, and microbe ratios will always be limited in capturing our lived experience, microbial census counts spotlight potential trouble areas. A healthy, resilient microbiome features complex biodiversity with representation across many phyla, classes, genera, and species. Unfortunately, modern lifestyle assaults chip away at microbial heterogeneity.

Chronic psychological stress, super-sanitized living conditions, persistent environmental toxins, and nutrient-deficient processed foods all diminish microbial richness over time. But through consciously curated diets, prebiotic and probiotic fortification, stress relief practices, and microbiome-friendly spaces, we can tip the balance back toward diversity. Of course, it's not just the number of different microbes but how they function together in a community that determines digestive health outcomes. Teams of researchers now track how disruption of stable equilibrium states in the microbiome lead to chaotic rollercoaster ecosystems, laying the foundations for inflammation and immune dysfunction.

The goal is to support reliable, resilient networks via lifestyle medicine - enhancing positive feedback loops among beneficial bugs. This fosters flexible, adaptive microbiomes that can handle occasional indulgences without veering too far into dangerous dysbiosis. Assessing current microbiome status provides the all-important baseline needed to judge the effectiveness of interventions while revealing one's core enterotype. Mapping out starting conditions helps shape realistic expectations around the timeframe to achieve balance.

Patience is key as it takes time to coax complex microbial communities toward equanimity. But armed with data and a deeper awareness of internal ecology, we can symbiotically guide our microbiomes toward harmonious diversity. Tracking indicators of gut health empowers us to consciously cultivate stability and resilience one microbe at a time through the choices we make each day. When cultivating a flourishing, biodiverse internal garden, we must move beyond probing the soil to actively planting the seeds of change. While assessing microbiome status provides an invaluable baseline, our daily lifestyle choices shape long-term resilience and equilibrium.

From our waking rituals to work habits and communal connections, nearly every aspect of modern life impacts our inner ecology. By consciously curating environments, behaviors and rituals that nourish microbial diversity, we can organically shift our inner terrain toward balance. The choices we make about where we spend our time wield tremendous influence over which microbes can thrive. While we rarely consider microbial impacts when visiting a friend's home or booking a hotel room, the truth is our bodies host very different bug bouquets based on the spaces we inhabit. Researchers now track how built environments teeming with certain bacteria, viruses, and fungi nudge our microbiome in different directions - for better or worse.

Places rife with mold, dust mites, chemical cleaners, and ambient air pollution promote inflammation, whereas natural spaces alive with phytoncides, negative ions, and diverse microbes enhance equilibrium. Even cultural practices like removing shoes before entering homes or building relational rituals around preparing and sharing food can sway the bug ballet. When we understand spaces as complex ecosystems that seed our microbiome, it empowers intentional choices for diversity. Seeking regular contact with lush green spaces alive in microbial synergies supports equilibrium, as does spending time in homes or hotels using plant-based cleaners. Maintaining moisture balance through proper ventilation and humidity modulation provides a canvas for microbial artistry. Displaying fermented foods like kimchi and kombucha adds inspiring diversity to kitchen counters while avoiding the overuse of disinfectants keeps things lively. Choosing restaurants that source organic produce from microbial-rich soils also boosts exposure to health-promoting microbes missing from the standard Western diet.

The spaces we inhabit daily demonstrate that cultivating external microbiome diversity fuels our inner resilience. Of course, beyond where we spend time, endless lifestyle factors shape our internal ecology over which we have agency; from how we move and sleep to what we apply on our skin and even who we kiss. Walking barefoot occasionally exposes feet to heme-gobbling Staphylococcus lugdunensis while also earthing ourselves into the soothing negative ions of Mother Earth. Getting adequate, deep sleep allows critical microbiome restoration and blood-brain barrier repair to unfold, supporting diverse signaling between the gut and mind. Regular massage with anti-inflammatory aromatherapy oils feeds skin commensals, which communicate with circulating T-cells to reduce whole-body inflammation. Something as simple as having a dog lick our face may seed oral and nasal passages with new microbial allies. When we remain mindful of how lifestyle variables shape our inner ecology, once-invisible choices suddenly surge into view. From preparing meals with fermented condiments to checking ingredient lists for emulsifiers and artificial sweeteners, our power expands through presence.

What we choose - or refuse - to put in and on our bodies ripples out to influence whole-body equilibrium. Eliminating microbiome-disrupting filler ingredients cultivates diverse symbionts, while intermittent fasting allows old microbial alliances to dissolve. Choosing natural body care products nourishes niche skin ecosystems as yoga flows massage internal tissues, shaping circulation and oxygenation patterns. By broadening our sphere of concern to include microbial impacts, our agency expands to consciously seed and feed diverse inner allies through daily lifestyle medicine. Of course, beyond physical environments and self-care rituals, the realm of relationships also demonstrates microbiome co-creation.

From couples whose kissing and sexual intimacy allow unique oral microbes to jump between partners to parent-child transmission dynamics shaping everything from immunity to anxiety risk, the truth is we cultivate inner resilience in communal concerts. Researchers now show how even brief exchanges of affection allow bugs specialized in digesting trace nutrients to find new microbial neighborhoods to flourish within. The more intimate the exchange through practices like kissing, oral sex or sharing food from the same plate expands the probability of successful migration between bodies. In essence, the greater the diversity of our social connections, the more allies find mixer events to circulate symbiotic gifts. This microbial mingling through intimacy likely evolved to enhance adaptability and equilibrium. Just as oxidative stress can develop if we exercise the same muscles daily while ignoring opposing groups, failing to diversify shared microbes may limit systemic flexibility.

It is often our closest bonds that shape foundational enterotypes, giving microbes specialized at training our developing immunity. But opening to new forms of spiritual intimacy provides an opportunity for the discovery of additional helper organisms. Attending communal meals, dancing in celebratory spaces and sharing physical touch at public gatherings allows once-isolated microbes to intermingle. By consciously curating diverse intimate exchanges, we seed resilience through opening to new symbiotic alliances. Those managing autoimmunity and chronic inflammatory conditions require more restraint when exchanging microbes. But for mostly healthy individuals, consciously mingling microbes through intimacy boosts adaptability, just as the exchange of diverse ideas between cultures expands perspectives and possibilities. The key is maintaining a balance between bonding with baseline symbionts through existing relationships while allowing some diffusion and exchange through a wider social interface. Too much isolation creates stagnancy and inflammation, whereas excessive openness risks instability.

Aim instead for gradual, incremental exposure by occasionally kissing new partners hello, sharing food off communal plates or exchanging handshake embraces at gatherings. Through expanding intimacy networks, the hidden handshake of microbial exchange diversifies the inner terrain. When we pull back the veil of invisibility, cloaking the microbial dimensions of lifestyle and relationships, our world reshapes. Suddenly, spaces glow with unseen ecological complexity, self-care practices take on new meaning, and the realm of touch becomes charged with hidden possibilities. By choosing to consciously shape lifestyle variables through the context of their microbial impacts, our power to cultivate a flourishing inner ecosystem expands dramatically. Like master gardeners tending an external oasis of diversity through controlling water flow, pruning plants and curating the mix of species, we patiently guide our bodies towards balance by the small choices we make each day. The road to microbial harmony involves gradually replacing diversity-depleting practices with equilibrium-enhancing rituals over time. As new research reveals surprising links between lifestyle variables and bug blossoming, our consciousness expands to note previously hidden connections.

By staying attuned to signals from our sensing selves and tuning into how subtle shifts in habits change digestion, energy, cravings and mood, we sensitively titrate lifestyle tweaks. With an orientation of exploration over perfection, leave no action too small or mundane when assessing the potential to support unique inner ecosystems. Just as dropping a stone in one corner of a pool sends ripples throughout, each tiny choice harmonizes with many to shape whole-body symphonies.

Chapter 10

The Vaginal Microbiome and Reproductive Health

The delicate balance of microbes in the vagina plays a critical role in defending against pathogens, maintaining pH, and ensuring overall gynecological health. When disrupted, it can lead to issues like bacterial vaginosis, yeast infections, and infertility. In this chapter, we'll explore the complex microbial ecosystem of the vagina and how it interconnects with reproductive wellness.

The vaginal canal hosts a dynamic community of microorganisms, including bacteria, viruses, protozoa and fungi. Lactobacilli, a type of friendly bacteria, predominates in most vaginas constituting over 70% of the microflora. The most common Lactobacillus inhabitants are L. crispatus, L. jensenii, L. gasseri, and L. iners. As lactic acid producers, lactobacilli acidify the vagina to a pH of around 4.5, making it inhospitable to foreign invaders. While each woman's vaginal microbiome is essentially as unique as a fingerprint, five community types have been identified so far. Four are Lactobacillus-dominated, but the fifth contains a diverse mix of anaerobic bacteria like Gardnerella, Atopobium, Prevotella, and more. This fifth group lacks protective lactobacilli and is linked with bacterial vaginosis (BV). BV arises when the number of good to bad bacteria dips, allowing the overgrowth of pathogenic microbes. In addition to bacteria, healthy vaginas also harbor viruses, yeasts like

Candida albicans, and microscopic protozoan creatures called Trichomonas vaginalis.

Mostly commensal, they help stabilize the vaginal ecosystem, only causing issues if their populations explode. The presence of semen during intercourse also impacts the vaginal microbiome by altering pH and microbial communities. While much remains undiscovered about the vaginal microbiome, researchers are steadily unpacking its role in gynecological wellbeing. Tracking it during pregnancy, menstruation, and menopause continues to illuminate how it interacts with hormonal shifts in the female body. Even the microbes in male partners seem interconnected with vaginal health. As more and more women have careers in STEM, the depth and reliability of research on women's health and biology will surely increase. For now, though, it's crucial to balance vaginal health.

To keep vaginal microbes balanced, reducing activities disruptive to lactobacilli is key. Practices like douching, smoking cigarettes, and using scented feminine products can disturb lactobacilli, allowing pathogens to colonize. Natural is best! Oral contraceptives and antibiotics also influence microbial populations, sometimes permitting yeast or BV-associated bacteria overgrowths. However, it's still perfectly safe to take oral contraceptives when you need them. Even diet, stress levels, and toxin exposures affect the nether region's ecosystems. Supporting colonies of protective lactobacilli is vital for defending against gynecological infections. Probiotic supplements containing Lactobacillus strains like L. crispatus, L. Jensen, and L. rhamnosus help increase lactic acid production and crowd out problematic bacteria. Prebiotic fibers that feed friendly flora are also beneficial. Getting enough vitamins A, C, and D enhances epithelial barrier integrity preventing unwanted microbes from invading vaginal tissue.

The composition of a woman's vaginal microbiome also impacts her reproductive capacity. Microbial imbalances are associated with infertility, recurrent miscarriages, and preterm birth risk. Restoring equilibrium in the vagina may optimize the chances of conception and carrying pregnancies to full term. Even a male partner's microbes affect reproductive success, highlighting the need for both members to nurture balanced microbial communities. But even what is known now provides women with actionable steps to care for their reproductive area's microscopic inhabitants. In turn, the tiny creatures inhabiting vaginal ecosystems play critical roles in gynecological wellness and fertility outcomes.

Getting pregnant relies heavily on a vaginal environment conducive to conception - namely, an acidic pH and plentiful colonies of Lactobacillus. As prolific lactic acid producers, Lactobacillus strains lower vaginal pH creating a hostile habitat for sperm-killing pathogens. Protective biofilms formed by these bacteria also guard sperm from damaging inflammation and immunological attacks as they journey through the vagina. Women with lower proportions of Lactobacillus species face higher odds of infertility and recurrent miscarriages.

Without sufficient lactic acid for acidification, vaginal pH rises, allowing Escherichia coli (E.coli) and other microbes hazardous to sperm to thrive. Biofilms also weaken, leaving sperm more vulnerable to the female immune system. Specific strains like L. crispatus prove especially protective - women with L. crispatus dominate communities and exhibit better implantation, conception, and birth rates.

The influence of vaginal microflora on fertility persists even through IVF treatment. Studies on women undergoing IVF found higher levels of L. crispatus correlated with better embryo implantation, while depleted Lactobacillus increased failure rates. Testing and addressing vaginal dysbiosis before IVF may boost outcomes for aspiring mothers struggling with infertility. Sexually transmitted bacterial, viral, and parasitic pathogens also disrupt conception and pregnancy capacities, often via inflammation. Gonorrhea and chlamydia inflict damage all along the female reproductive tract, sometimes permanently scarring fallopian tubes. Trichomoniasis protozoa produce toxins extremely hazardous to embryo development. HPV transmission introduces viruses that can trigger cervical cell abnormalities. By acidifying the vagina and occupying epithelial niches needed by STDs for colonization, Lactobacillus offers powerful protection against these reproductive threats.

Women with Lactobacillus-deficient vaginas face exacerbated risks of contracting gonorrhea, chlamydia, trichomoniasis, bacterial vaginosis, and more if exposed. Preserving ample Lactobacillus colonies minimizes STD access to reproductive tissues, safeguarding fertility. A mother's vaginal microbes critically impact the pregnancy journey at every step. Imbalanced, inflammation-triggering microflora impairs embryo implantation while also increasing risks of preterm birth later on. Studies consistently connect pregnancies resulting in preterm delivery with a lower abundance of Lactobacillus species and the dominance of other anaerobic bacteria. The absence of Lactobacillus allows pathogenic communities to flourish, eliciting inflammation and driving early births. Resulting changes in immune factors and mucus production in the vagina may also contribute. Interventions restoring Lactobacillus equilibrium demonstrate some efficacy in preventing premature labor - especially in women with a prior history of preterm births.

Later in pregnancy, ascending infections facilitated by depleted Lactobacillus also endanger fetal development. Urinary tract infections contracted by expectant mothers readily spread to the intra-amniotic space if protective vaginal biofilms falter. Unchecked, these fetal infections spark inflammatory cascades that can trigger premature contractions, early membrane rupture, sepsis, and more. With reproductive and pregnancy health so intimately interwoven with vaginal microenvironments, optimizing microbial communities is vital. Testing for imbalances like bacterial vaginosis and facilitating Lactobacillus flourishing helps set the stage for conception while securing gestational success once pregnant. Probiotic therapy is one of the most powerful tools for rebuilding protective Lactobacillus colonies. Oral probiotics help restore equilibrium from within through microbial seeding of the gut. Strains like L. rhamnosus and L. reuteri found in supplements readily migrate to the vagina upon ingestion, reinforcing populations there.

Local vaginal probiotics also recolonize Lactobacillus through direct injection in the vagina. Applications of gels, suppositories, ovules and more containing strains like L. crispatus, L. jensenii, and L. gasseri resupply levels depleted after antibiotic use, pH disruption, douching, and infections. Using both oral and local routes synergistically helps regrow robust biofilms of microbes instrumental in the conception and gestational defense. In addition to therapeutic courses restoring existing imbalances, prophylactic probiotic use also wards off future threats. Maintaining abundant Lactobacillus through daily oral and periodic vaginal probiotics fortifies microbial protection needed for reproductive wellness. Just like guarding intestinal health with consistent probiotics, safeguarding the vagina with Lactobacillus leverages microbial might to shield maternal-foetal wellbeing. Looking at the reproductive microbiome also further informs probiotic selection. Insights like these help tailor probiotic remedies toward specific reproductive protections women need along the pregnancy journey. On top of restoring.

Lactobacillus abundance directly, supplying prebiotic fibers that fuel friendly flora also cultivates fertility-promoting conditions. Plant polysaccharides like inulin, linen seeds, chicory, and more get directly metabolized by Lactobacillus into lactic acid that lowers pH. This makes the vaginal environment more hostile to STDs and sperm-destroying pathogens attempting colonization. Prebiotic fibers also enhance nutritional capacity, supporting heavier Lactobacillus biofilms with more robust adhesiveness. Well-fed, satiated colonies of microbes like L. crispatus and L. jensenii cling onto vaginal mucosal terrain better-obstructing areas needed by invaders to establish infection. Denser lactobacilli biofilms also filter out more incoming threats, blocking passage upwards into the uterus. Incorporating prebiotic-rich foods like garlic, onions, lentils, and bananas or supplements into diets gives protective vaginal microbes the sustenance they need.

This further multiplies colonies, creating a stronger hedge against fertility-disrupting pathogens. Addressing any nutritional deficits that constrain microbial growth removes limitations, allowing Lactobacillus communities to flourish fully. The interconnected microbial ecosystems spanning the gut, vagina, and even oral cavity also underscore the value of whole-body prebiotic nourishment. Researchers now recognize many Lactobacillus strains originating in the vagina first seed there from intestinal populations.

Ensuring plentiful sustenance through dietary prebiotics means more Lactobacillus can migrate between linked microbial niches. Beyond directly manipulating microbial communities themselves, supporting the environmental conditions needed for equilibrium also proves important for vaginal wellness. Avoiding factors that disrupt microbiomes and promote inflammation minimizes collateral damage, enabling Lactobacillus to recover more readily. Concurrently, lifestyle measures that resolve inflammation, enhance detoxification, and improve antioxidation help remedy tissue terrain. Anti-inflammatory nutrients like omega-3s and curcumin combined with glutathione boosting through NAC, milk thistle, and vitamin C facilitate the healing of damaged mucosal lining. Supporting liver detoxification also clears away harmful estrogen metabolites implicated in hormonal imbalances disrupting vaginal flora. Leveraging the power of whole food, plant-based diets additionally nurture the vaginal landscape. Micronutrients from organic vegetable intake balance hormones while decreasing inflammatory triggers from meat, dairy and refined carbs. Eliminating blood sugar spikes that dysregulated immunity further relieves downstream inflammation manifesting as vaginal irritation.

Empowering women with actionable steps that reinforce the microbial and environmental pillars underlying vaginal wellness gives Lactobacillus and other protective microbes the best chance of thriving. In turn, balanced, robust vaginal microbiomes unlock the gateway to optimizing female fertility and securing safe passage through pregnancy for hopeful or expecting mothers.

Chapter 11

Skin Microbiome: Barrier Protection and Immunity

The skin is the body's largest organ, covering about 20 square feet in adults. Far from being an inert barrier, skin hosts a diverse ecosystem of microbes that interact with the immune system to influence the health of the whole body. The types and balances of microbes making up the skin microbiome can profoundly impact skin disorders, allergies, autoimmune diseases, and infection risk.

The skin microbiome refers to the trillions of bacteria, viruses, fungi, and other microorganisms living on the surface of our skin and in skin crevices and folds. Most reside in the skin's superficial layers and are collectively referred to as "resident microbes." The skin microbiome varies significantly from person to person and even from one site on the body to another, depending on moisture, sebum levels, and exposure to environmental microbes. The majority of skin microbes are harmless or even beneficial "commensals." These aid immunity by taking up space and resources so pathogens cannot establish, producing antimicrobial compounds, and interacting with skin cells to strengthen the barrier.

A healthy, balanced skin microbiome checks bad microbes through this "colonization resistance." Imbalances in the skin microbiome are linked to disorders like acne, eczema, psoriasis and chronic wounds. The skin microbiome develops slowly after birth, shaped by the mode of delivery, environmental exposures, and contact. Babies born vaginally accumulate microbes from their mother's birth canal. In contrast, C-section babies initially have skin resembling their mothers' skin but acquire more microbes from their surroundings over the first year. This early colonization lays the foundations for the adult skin microbiome and immunity.

The composition of microbes on adult skin depends heavily on the local habitat. Moist areas like armpits and feet host vastly different bacteria than dry areas. Body sites also differ in pH, sebum and sweat production, and environmental exposure to microbes. The interactions between resident microbes themselves also help determine who thrives where. Understanding skin as an ecosystem helps explain why one-size-fits-all skincare fails. Many immune cells reside in skin layers, equipped to detect and destroy invaders. Specialized cells called Langerhans patrol for threats. Recent findings show the skin microbiome trains these cells early in life by exposing them to commensal microbes. This contact enhances the flexibility and function of Langerhans cells through an epigenetic process — altering what genes can be turned on or off. Such reprogramming by good bacteria enables calibrated immune responses to pathogens instead of overreactions causing chronic inflammation.

This emerging science reveals that the skin microbiome does far more than occupy space — it actively shapes local and body-wide immunity via direct contact with skin and immune cells. Associations exist between skin microbiome disruption and immune issues like food allergies, asthma, arthritis, Multiple Sclerosis, Type 1 Diabetes, and more. A better understanding of these connections could inform the prevention and treatment of many inflammatory conditions starting in childhood. Recent studies highlight the ability to select beneficial strains from the skin microbiome to strengthen the skin barrier, modulate inflammation, and hinder pathogens when applied topically. Translating findings about the skin immune-microbe axis into microbiome-targeted treatments holds great promise for diverse skin disorders plus whole-body immune health. Fortunately, our skin microbes are not fixed at birth but change across our lives with exposures and habits. Small intestine bacterial overgrowth (SIBO) involves inappropriate growth of bacteria in the small intestine. SIBO can cause bloating, abdominal pain, diarrhea and malnutrition through malabsorption of nutrients. Some studies show a higher prevalence of skin conditions among SIBO patients, likely resulting from systemic inflammation or microbial translocation from gut to blood to skin.

Treatments improving SIBO often resolve associated skin complaints, highlighting the gut-skin microbiota axis. In recent decades, inflammatory skin disorders like acne, eczema, and psoriasis have become more prevalent in developed nations. These often chronic conditions significantly reduce quality of life and currently have no medical cure, only management. Mounting research implicates the loss of key beneficial microbes from modern lifestyles as a contributor. Studies consistently relate low skin microbiome diversity to common skin disorders or irritation. Such findings underlie appeals from scientists for radical public health changes encouraging "biodiversity exposure" to boost immune resilience through healthy skin colonization.

Our modern obsession for cleanliness disrupts the transmission of essential microbes, but we co-evolved, requiring that very transmission. Thankfully, probiotics and prebiotics as supplements and functional foods now allow convenient microbiome enrichment lacking in sanitized lifestyles. Food components like beta-glucan, antioxidants, and plant polyphenols also support healthy skin microbial balance as emerging prebiotics. Ongoing research continues to illuminate how skin microbial inhabitants interact with local cells and our internal physiology. These communities mature along different trajectories based on mode of birth, environment, hygiene practices, contact with animals, and diet. Loss of microbial exposures key to immune education underlies surging inflammatory problems. Yet studies confirm simple regular contact with green space and pets still strongly enriches the skin microbiome, even for children and adults.

Our choices can nurture microbial friends, uplifting whole-body wellness through a balanced and biodiverse skin ecosystem. One proposed mechanism linking gut and skin dysfunction is the migration of pro-inflammatory TH17 cells. These immune cells develop in excess with some forms of gut inflammation and can travel via blood to distant sites like the skin. Here, they drive inflammation-flaring conditions like psoriasis and eczema. Studies confirm resolving intestinal inflammation reduces circulating TH17 cells and controls flare-ups of skin issues.

These findings inform combined treatments for optimal control of inflammatory disorders manifesting in dual areas. The influential nature of this gut-skin axis indicates holistic care for skin complaints should assess gut health and vice versa. Diets nurturing a balanced gut microbiome help resolve systemic drivers of skin inflammation originating internally. Reducing the intake of high glycaemic foods, emulsifiers and ultra-processed fare while increasing fiber-rich plant foods, ferments and collagen protects both organ systems. Lifestyle approaches like daily stress relief and circadian rhythm balance also benefit the microbiomes of our insides and outsides.

In recent decades, inflammatory skin disorders like acne, eczema, and psoriasis have become more prevalent in developed nations. Contact with pets, livestock, green space, and biodiverse environments enriches the human skin microbiome and correlates to reduced risk for allergic diseases. Benefits peak in early childhood, with some transferable to mothers during pregnancy. Links also exist between cesarean delivery, formula feeding, antibiotics, and household disinfectant use with higher rates of childhood allergies and associated skin issues like eczema. Such findings underlie appeals from scientists for radical public health changes encouraging "biodiversity exposure" to boost immune resilience through healthy skin colonization. Our modern obsession with cleanliness disrupts the transmission of essential microbes, but we co-evolved, requiring that every transmission.

Thankfully, probiotics and prebiotics as supplements and functional foods now allow convenient microbiome enrichment lacking in sanitized lifestyles. Food components like beta-glucan, antioxidants, and plant polyphenols also support healthy skin microbial balance as emerging prebiotics. Ongoing research continues to illuminate how skin microbial inhabitants interact with local cells and our internal physiology. These communities mature along different trajectories based on mode of birth, environment, hygiene practices, contact with animals, and diet. Loss of microbial exposures key to immune education underlies surging inflammatory problems. Yet studies confirm simple regular contact with green space and pets still strongly enriches the skin microbiome, even for children and adults. Our choices can nurture microbial friends, uplifting whole-body wellness through a balanced and biodiverse skin ecosystem.

Fortunately, research reveals that targeted lifestyle and nutritional approaches can improve skin dysbiosis by working with the microbiome. Regular biodiverse nature exposure encourages healthy, diverse colonization that is more resistant to pathogens. Probiotic skincare products deliver beneficial microbes, while prebiotic compounds feed and fortify existing residents. Inside the body, probiotic, prebiotic, and phytonutrient daily supplementation builds microbiome and immune resilience from the inside out. Studies confirm that oral probiotics positively impact the skin in multiple ways. Some strains reduce skin inflammation, increase ceramide production for stronger barrier function, and elevate antioxidant status. For acne, certain species curb excessive sebum production while others hinder P.

Acnes growth or neutralize its inflammation-provoking chemicals. Amazingly, drinking good bacteria can calm inflammatory skin disorders by modulating systemic immune function. Matching strains to pathways influencing each condition allows customization for optimal therapeutic effects. Besides probiotics taken internally, recent research shows promising impact from topical skin applications. Strains like Vitreoscilla filiformis, Roseomonas mucosa, and Lactobacillus paracasei demonstrate anti-inflammatory, barrier-enhancing, and antimicrobial benefits during studies on skin disorders like eczema, rosacea, wounds, and burns.

The notion of "skin probiotics" continues gaining traction for diverse dermatological applications. Their effects likely involve strengthening good microbes in the skin microbiome since transient surface contact alone seems inadequate to shift resident communities long-term. Nonetheless, the immuno-modulatory capacity of microbes without needles, pills, or side effects has researchers eager to uncover strains and serums that heal. Akin to gut microbiome diversity, an increased variety of bacteria living on the skin is strongly associated with health. Having robust and balanced skin flora makes colonization by pathogens less likely through simple spatial exclusion. More diverse biofilms may also resiliently withstand perturbation. Researchers propose actively "re-wilding" human skin through emollients, cleansers, and moisturizers, transferring healthy microbes. Be careful not to remove all of the bacteria or cells on your skin through chemical peels or other aggressive skincare routines. Such topical probiotics introduce novel microbial strains while prebiotic compounds nourish existing inhabitants.

The indispensable role of diet in building microbiome diversity from within also applies to skin health. Since inflammation represents the common thread connecting most skin disorders, anti-inflammatory foods help extinguish fires. Primarily plants, these cooling foods also act as prebiotics — preferential nourishment for friendly skin bacteria. Rich in antioxidants, fiber, plant polyphenols, minerals, and omegas, vegetables, herbs, teas, seaweeds, fruits, and ferments feed microbes, producing metabolites beneficial for local cells. A rainbow diet taxes pathogens while nourishing peaceful commensals for balanced cohabitation.

Elimination of inflammatory modern foods also allows the microbiome and immunity to recalibrate. Top offenders include high glycemic foods, emulsifiers, processed oils, food additives, and ultra-processed fare. Reducing animal foods, gluten, and dairy can relieve those with sensitivities as hidden drivers of inflammation. Minimizing intake of these while maximizing an array of prebiotic plants and ferments helps rebalance the gut and skin ecosystems. Tracking reactions to reintroducing eliminated foods helps personalize an optimal anti-inflammatory, microbiome-friendly diet.

While professional support has merit for restoring microbiome balance in cases of dysbiosis, daily lifestyle choices make or break skin health. Spending abundant time outdoors immersed in biodiverse nature reliably enriches skin flora through simple contact. Regular moderate exercise benefits the skin microbiome through circulation and immune effects. Stress management is essential since psychological strain chronically inflamed the skin. Gentle cleansing, moisturizing, and protection from chemicals, allergens, and irritants prevent disruption of delicate biofilms. Probiotic skincare and fermented topicals show particular promise for nourishing skin symbionts. Kindness to our microbial allies through conscientious daily habits empowers them to care for our largest organ in return kindly.

Chapter 12

Oral Microbiome: Gateway to Whole Body Health

Good oral hygiene does more than prevent tooth decay and gum disease. Research shows that those with periodontitis have a higher risk of cardiovascular disease. The inflammation from infected or irritated gums triggers white blood cells to release pro-inflammatory chemicals.

These end up circulating systemically and damaging arterial walls. Bacteria from gum tissue can also enter the bloodstream directly and promote clot formation. Vivian, 54, did not realize her inflamed gums were connected to the minor heart attack she experienced. "I'd had bleeding gums for years and put off seeing a dentist for too long," she admits. "Now I know failing to care for my teeth and treat early gum disease may have paved the way for bigger health issues." Studies indicate regular dental visits and plaque removal may lower cardiovascular risks significantly.

Similarly, acute necrotizing ulcerative gingivitis (ANUG) is a severe gum inflammation requiring urgent care. Also called "trench mouth", it can spread between people via saliva. ANUG thrives in nutrient-deficient conditions when oral hygiene habits slip. New research shows that pregnant women with an impaired oral microbiome are likelier to pass gingivitis-causing bacteria to their babies. This puts their little ones at higher risk for respiratory infections, ear infections, and other ailments stemming from inflammation coming from the mouth. "We originally thought nursing infants got exposure only after their first teeth erupted," explains pediatric dentist Dr. Judith Valentin.

"But even for the youngest of babies, plaque-forming bacteria from a mother's mouth can be transmitted and thrive in different parts of an infant's body if immune defenses are still developing." Catching gum inflammation early can halt the overgrowth of harmful microbes before they spread further in the body or to others.

More and more data links periodontitis and related inflammation to cognitive decline issues in aging populations. Though not causing Alzheimer's directly, poor oral health appears connected to the progression of amyloid plaques and tangles in the brain that lead to dementia symptoms. Just as other chronic inflammatory states promote neurodegeneration, so can the long-term presence of plaque buildup and gum irritation.

"It's likely the pro-inflammatory molecules released due to damage or infections in the mouth work their way to the brain via the bloodstream," Dr. Poonam Jain hypothesizes. "Once there, they may speed up inflammatory pathways underlying memory loss and other changes." Carlos, 67, thinks twice about skipping dental cleanings after his dentist cautioned him about the newly discovered linkages. "No one wants to deal with dementia down the line if you can help it. Getting my plaque cleaned regularly might help me keep a sharp mind as I age."

The body suffers greatly when the dental region becomes diseased and imbalanced by harmful bacteria. Experts now rank poor oral hygiene as a major modifiable risk factor for myriad illnesses. But a healthy, well-cared-for oral microbiome sets the stage for good nutrition, a robust immune system, and reduced systemic inflammation. Optimizing our oral flora diversity through conscientious home and professional care pays untold long-term dividends for oral and whole-body wellness. Protecting oral ecological balance serves our health by safeguarding digestive processes, cardiovascular function, mental acuity, and more. As the gateway to our microbiome universe, perhaps it's time we start tending to our teeth and gums with the same care we show our gut.

Chapter 13
Infant Microbiome Development and Diseases Risk

The process of microbial transmission begins well before birth, with populations of lactobacilli and other bacteria in the maternal vaginal and intestinal tracts supplying a ready source of pioneer species. As many as 100 trillion microbes may coat the surface of the placenta, though few penetrate deeper fetal tissues kept sterile by the amniotic sac and immune factors.

When the water breaks, however, all bets are off - the formerly protected fetus becomes bathed in a stew of beneficial yet foreign bacteria. For most of human history, this microbial baptism occurred via vaginal delivery, allowing infants to acquire species uniquely adapted for early gut and skin colonization. In recent decades, C-sections have become much more widespread, altering this finely-tuned process in ways we are just beginning to grasp. At the same time, cavity pressures and contractions help seed the baby's gut with maternal fecal microbes - key symbionts like Bifidobacterium and Bacteroides colonize within days or weeks when acquired this natural way. Together these vaginal and gut microbes fulfil pioneer species roles, staking early claims on skin and intestinal real estate to promote homeostasis.

By contrast, C-sections allow doctors to bypass the vaginal canal, limiting exposure to this ready reservoir of specialized bacteria. Multiple studies show C-section babies harbor altered gut, skin, and even brain microbial profiles, lacking vaginal lactobacilli and showing reduced Bifidobacterium colonization for months after birth. Rates of other pioneer gut microbes like Escherichia are also higher compared to vaginally-born infants in the first 3-6 months of life.

While C-sections are sometimes medically necessary, avoiding them allows the natural seeding of lactobacilli and other pioneer species in the earliest moments of life. When surgery is required, practices like vaginal seeding and skin-to-skin contact with the mother provide at least partial injection and should be encouraged. Promoting breastfeeding as early as possible also helps drive more rapid Bifidobacterium colonization in C-section infants - human milk oligosaccharides act as prebiotics feeding this beneficial genus. In any birth scenario, a diversity of lactobacilli, Bifidobacterium, and key co-colonizers in the first days and weeks of life lay vital ecosystem foundations. As pioneer bacteria stake their claims, they pave the way for successive waves of colonization towards a mature, balanced state - but only if supported by further healthy exposures.

Birth kicks off gut microbial acquisition but is just the opening act - early feeding choices such as breast vs formula feeding represent the next critical juncture for populating our inner ecosystem. Regarding microbial exposures, not all nutrition sources are created equal - and none compare to the co-evolutionary magic of mother's milk. Breastfeeding has myriad known benefits for infant health, including reduced infections, allergies, obesity, and other inflammation-driven conditions. In recent years, these protective effects have been tied to breast milk's impact as a "microbiota transplant" from mother to child. Beyond simple nutrients, human

milk transfers an array of pioneer bacteria, health-promoting short-chain fatty acids, and specialized prebiotic compounds feeding our infant gut symbionts.
All breast milk naturally contains bacteria from the mother's gut and skin microbiome, albeit at lower levels than common probiotic foods like yogurt.

Early studies using culture-based techniques detected low overall biomass in breastmilk, consisting mostly of staphylococci and streptococci from the skin alongside gut microbes like lactobacilli and bifidobacteria. The advent of genetic sequencing revealed just how undersampled milk microbiota was via classic techniques - while biomass remains low, diversity is far higher with over 200 species identified. Proteobacteria, Firmicutes, and Actinobacteria dominate, with Bifidobacterium, Staphylococcus, and Streptococcus common alongside Faecalibacterium, Lachnospira, and other gut symbionts. Researchers now think much of this community originates in the maternal intestinal tract, secreting bacteria into ductal tissues lining the breast that avoid washout from milk flow. Major differences in diversity and composition exist between colostrum, transitional, and mature breast milk.

Colostrum produced in the first days after birth has higher bacterial density and biomass, dominated by skin commensals like Staphylococcus and Streptococcus. After lactogenesis II kicks in around 3 days postpartum, transitional breastmilk emerges with increased bifidobacteria and lactobacilli. By one month postpartum, a mature milk microbiome rich in anaerobic gut microbes like Bifidobacterium breve, B. infantis, and B. longum is well established. The stability of this community depends greatly on feeding mode, with direct breastfeeding promoting more resilient, diverse profiles than pumped milk. Freezing and storage also drastically alter the milk microbiome, reducing bacterial biomass and diversity - feeding fresh whenever possible provides ideal exposure. Beyond live microbes, breast milk delivers a wide array of prebiotics and immunomodulators, feeding our beneficial bacteria and shaping their evolutionary trajectory. Exclusively breastfed infants show select strains of bifidobacteria enriched in their gut biomes able to digest specialized HMOs like lacto-N-fucopentaose and lacto-N-neotetraose. These same symbionts digest complex sugars in milk, producing abundant short-chain fatty acids that support immune and gut barrier development.

In concert with pioneer species acquired during vaginal birth, this unique nutrition helps bifidobacteria and key co-colonizers dominate the breastfed infant gut through six months of life. Formula feeding similarly lacks coevolved nutrition and microbial exposures. While some bifidobacterial strains can break down prebiotic fructooligosaccharides in formulas, most are poorly adapted without human milk. Multiple studies show increased potential pathogens like Clostridium, Staphylococcus, and E. coli in formula-fed infant guts alongside reduced overall bacterial richness and diversity. Ratios of Firmicutes to Bacteroidetes also shift compared to breastfed controls, reflecting more pro-inflammatory gut states frequently dysbiotic into childhood.

Altered digestive capacity represents one local outcome of delayed or insufficient colonization. Infants fed lower diversity diets or lacking pioneer symbionts show reduced long-term enzymatic capacity to break down complex carbohydrates. This hampers nutritional status and growth into childhood while increasing inflammatory metabolites and gut permeability. But beyond digestion itself, gut microbes intimately communicate with intestinal tissues as they develop - disruptions to signaling pathways in early life permanently alter barrier integrity, immune function, and enteric nerve wiring with systemic consequences. Infants fed formula rather than breastmilk, for instance, show lasting increases in inflammatory biomarkers, weaker gut barrier function, and altered vagus nerve connectivity to the brain. The most well-studied outcome is heightened allergy and asthma risk after low-diversity infant exposures.

Lacking old friends to train early immunity, children fed formula instead of breastmilk see twice the incidence of eczema by age 1 year, along with worse wheezing and allergic reactions through the toddler years. Multiple studies trace this immune dysregulation back to insufficient seeding with beneficial microbes needed to prime tolerance development in utero and early postpartum. Beyond allergies and asthma, diabetes and obesity incidence are closely tied to infant feeding modes and antibiotic exposures as well.

Nearly twice as many exclusively formula-fed children become overweight or obese by age 3 years old, while also seeing higher fasting blood sugar and diabetes risk into adolescence and adulthood. Mouse models demonstrate gut microbes directly mediate at least some of this metabolic disease shown in retrospective human studies. Exactly how early-life dysbiosis transfers such lifelong effects remains under investigation, but compelling pathways exist. Altered pioneer species colonization changes the expression of host genes regulating digestion, immunity, and intestine-brain-endocrine crosstalk early on, skewing developmental blueprints toward disease. Sustained low diversity also hampers tolerance at sensitive training windows, allowing pathogens like Staphylococcus and Klebsiella to dominate communities and trigger inflammatory cascades.

Epigenetic modulation also likely plays a role - differences in DNA methylation and histone wrapping depending on milk feeding are now apparent. Formula-fed infants show distinct CpG methylation patterns in over 1500 genes, mainly regulating immunity and metabolism, compared to exclusively breastfed children. Gut microbes and their metabolites may directly alter these epigenetic tags during sensitive windows, sustaining inflammatory signaling long-term. Together, these intertwined mechanisms ensure early life microbial exposures alter risks of some diseases for a lifetime, no matter what interventions come later.

But if the first months of life harbored such sway over disease odds alone, microbiome research would merit only passing interest - after all, few recall their infant feeding history in detail by adulthood. What makes the microbiota so captivating and promising is that to some degree, these development processes remain plastic for decades more as new exposures continuously shape the balance of bugs and hosts.

Despite heightened sensitivity in the first months postpartum, gut ecosystems stay somewhat malleable as exposures throughout childhood and adolescence continue to guide maturation toward optimal diversity and function. A healthy diet, lifestyle, and environmental inputs help reinforce disease-protective communities acquired via ideal birthing and infant feeding wherever possible. As solid foods become a larger part of children's fare from ages 1 to 3 years, microbiome diversity escalates dramatically thanks to varied polysaccharides entering the gut from plants. Key genera missing in children fed only milk and processed snacks get opportunities to establish when exposed to whole vegetables, tubers, legumes, and fermented foods at this age.

High fibre fruits in particular, help enrich communities rapidly between ages 2 to 5 years old based on eating frequency. Apples, bananas, and citrus provide a range of soluble and insoluble fibers supporting Saccharibacteria, Bacteroides, and other genera missing in Western children more reliant on refined grains and animal products early on. Through the elementary school years, from ages 6 to 11, gut microbiomes show another large gain in diversity and stability if exposures remain consistently healthy. As children go to school with other children, there is huge exposure to new bacteria. This explains why there are regular colds and flu in schools and homes for what seems like almost the entire year! Regular phytochemical inputs from colorful produce, nuts, seeds, and low-gluten ancient grains feed communities processing complex carbohydrates and plant metabolites.

Fermented foods further help establish robust acid-loving lactic acid bacteria in later childhood, where continued exposures occur. Another key input starting around ages 4 to 7 years emerges in the form of pet ownership, which shows huge microbial community impacts when acquired in early life. Children exposed to furry pets like dogs and cats from an early age harbor distinct gut and skin microbiomes by age 11, enriched in environmental microbes shared with their four-legged friends. Frequent animal contact also aids immunity - children in pet-owning households see up to 30% fewer upper respiratory tract infections in childhood overall.

But as with other sensitive windows, missing health-promoting exposures at key ages can reduce chances of nurturing optimal microbiomes long-term. School-aged children eat plenty of sweets and ultra-processed snacks but few vegetables see lasting delays in the colonization of plant-loving microbes.

Mass antibiotic exposures for repeat ear and throat infections forever wipe out select symbionts in some children as well. And moving from rural to urban environments, stressed microbes including fungi and viruses, also dwindle by tween ages with lasting impacts on allergies and mood issues tied to microbial inputs. While another marked shift towards more adult-like configurations happens from ages 12 to 17, corresponding to puberty, inputs during this time continue subtly shaping long-term diversity and function, too. Teens who eat well, stay active, keep stress balanced, and avoid antibiotics whenever possible end up with the most robust, beneficial communities by adulthood. But even for those lacking certain pioneer species long-term, supplemental prebiotics, probiotics, phytochemicals, and healthy lifestyle exposures through adolescence provide the potential for some continued maturation and disease protection over time.

Chapter 14

Environmental Microbiomes: From Home to Hospital

Step into any human-made structure, from an office high-rise to a suburban home, and you enter an ecosystem alive with microbes. Indoor spaces have their unique microbial communities that profoundly impact human health, for better or worse. With people in developed countries spending roughly 90% of their time inside buildings, understanding how our indoor microbiome exposure influences wellbeing and disease risk has become a major research priority.

The microbial profile of a home varies based on numerous factors like geographic location, building materials, number of occupants, and lifestyle habits. Carpets, HVAC systems, kitchens, and bathrooms harbor abundant and diverse microbes. However, moisture issues and mold create less benign microbial growth. Excess indoor humidity encourages mold species, triggering respiratory issues, cognitive dysfunction, and weakened immunity. Those living in water-damaged buildings have demonstrated nose, throat, and skin inflammation and higher cardiovascular disease markers. Balanced indoor microbiomes support health, while imbalanced, molded ecosystems increase inflammation and disease risk.

Modern office buildings allow people to work together nearby, also enabling microbes to easily spread through indoor air, surfaces, and human contact. Air sampling studies demonstrate substantially higher microbial content in offices compared to outdoor air. Office design can greatly impact transmission rates.

Open floor plans and shared workspaces have more microbial exchange occurrences than private offices. Researchers in Amsterdam found that when co-workers share physical items like phones and keyboards, their gut and mouth microbial profiles also become more similar. Unsurprisingly, the office kitchen areas show the highest density of surface microbes. Fomites like copy machines, water coolers, and stair railings also harbor varied bacteria, viruses, and fungi. Think about how often they are touched and used versus how often they are cleaned! Those working at desktop computer stations demonstrate unique skin microbiomes on their hands and wrists.

While mostly harmless skin commensals dominate these areas, some disease-causing strains have also been identified. Overall, office microbial diversity does not always guarantee health benefits. Viruses tend to spread rapidly through densely occupied indoor work areas. However, strategic office planning to control occupant density and surface decontamination protocols could support microbial balance. Hospitals work diligently to eliminate pathogens yet still battle extremely high infection rates. With constant traffic of new patients and staff, nearly all hospital surfaces house complex microbial networks.

Hospital air also contains various bacteria, especially in inpatient rooms. One concerning finding is that the microbiomes of hospitals often contain antibiotic-resistance genes. Bacteria isolated from clinic floors, medical tools, and nurses' uniforms demonstrate resistance to major antibiotics. This likely results from the heavy use of antibiotics driving microbial adaptation. Of highest risk are ICUs and surgical suites, where microbial density can approach 10,000 times that of outdoor air samples. These areas still demonstrate significant bacterial loads despite rigorous disinfection routines, placing vulnerable patients at risk.

However, hospitals also rely on beneficial microbes to treat serious infections. Fecal microbiota transplant has shown profound success in treating stubborn C. difficile gut infections. As medical understanding of microbiomes expands, similar microbial therapies will likely increase. Ultimately, hospitals sit on the frontline of the evolutionary battle between infectious microbes and medical science. Their walls contain microscopic dramas with the power to save lives or take lives. Continued research, better architectural planning, and microbial community analysis will help optimize hospitals' life-giving ecological networks.

We spend almost all our time moving between indoor ecosystems that shape our microbial exposures and related health impacts. Increased focus on optimizing household, office, and hospital microbiomes could significantly influence public health and reduce disease burdens. Promoting microbial diversity through eco-friendly building materials, smart architectural choices, and microbiome-focused healthcare policies offers practical disease prevention strategies. As research advances, buildings may undergo "microbiome testing" along with structural inspections before tenant occupancy. We stand to benefit greatly from properly balancing the indoor ecosystems that surround us.

Chapter 15 The Future of Microbiome Medicine

Microbiome medicine holds incredible promise to transform how we understand, prevent, and treat illness. Recent advances allow us to glimpse the vast potential impact microbiome insights can have for human health. This chapter will explore cutting-edge microbiome research and envision how we may harness microbiome science to improve well-being.

We'll journey into a future where personalized probiotics, precision editing of gut microbes, customized diets based on individual microbiome profiles, advanced microbial testing, ecosystem medicine approaches, and microbiome-targeted drugs and treatments are part of standard medical care. Exciting microbiome findings could also help reduce disease transmission, assist immuno-oncology, and enable tailored mental health interventions.

Microbiome research has birthed a radical new paradigm for conceptualizing health and illness. Rather than viewing disease as an invasion of external pathogens, the microbiome lens sees health as an emergent property of complex microbial/human interactions. Imbalances in this two-way dialogue, not just pathogen invasion, can trigger systemic inflammation underlying chronic illness. This paradigm views humans as rich ecosystems where boundaries between "self" and "other" blur. Our bodies contain roughly equal numbers of human and microbial cells, with microbial genes outnumbering humans by 150 times. We are complex communities of diverse micro- and macro-organisms living interdependently. Health emerges from nurturing mutually beneficial relationships. Much chronic disease may arise when this community breaks down, creating inflammation that sabotages systemic functioning.

Factors like stress, toxins, poor diet, and antibiotics can disturb the body's Internal Ecosystem, reducing diversity and resilience. Rebalancing community dynamics offers paths for restoring health. One microbiome frontier involves developing hyper-personalized probiotic formulas tailored to an individual's unique gut microbiota. Today, most probiotics contain general bacterial strains considered broadly beneficial. But our deeply individualized microbiomes mean no one-size-fits-all solution exists. Precision probiotics would instead culture targeted microbial communities from healthy donors compatible with a recipient's microbiome. These custom ecologies could then rebalance the gut optimally. Early research successfully treated C.

difficile infections by transplanting tailored, complimentary collections of gut microbes from donors to patients lacking health-promoting flora. Further innovation could bioengineer personalized probiotic formulas cultured specifically to fill deficiencies within a client's microbiome and digestive health needs. Microbe-mapping services like Viome already analyze microbiome profiles seeking pathogenic strains correlating with health risks. In the future, such testing could guide the design of made-to-order probiotic combinations, restoring microbial balance. Companies are also working on synthetic microbiomes mixing beneficial microbes, showing promise for personalized therapies. Ongoing innovation may also enable on-site cultivation and harvesting of microbiota within medical settings. FMT treatment could utilize microbial mixtures sourced from the patient during colonoscopy prep. This would reduce infection risks associated with using third-party donors in DIY at-home FMT attempts. Custom diets may similarly emerge based on specific microbiome profiles. New companies offer testing analyzing how particular foods uniquely impact an individual's microbiome.

These insights allow crafting eating plans to nurture optimal gut flora. Combining such diets with personalized probiotics could profoundly influence wellbeing. Metagenomic testing analyzing microbiome genes may also inform future probiotic development. Distinct microbial genes drive key activities like short-chain fatty acid production and inflammation modulation. Metagenomic insights could guide probiotic design targeting genetic capacities missing within compromised microbiomes. This field remains young, but further research may uncover disease-associated gene patterns enabling personalized probiotic therapies. Beyond microbiome mapping and precision pairing, new biotech could enhance probiotics' effectiveness dramatically.

For example, advances in nanoparticle delivery systems show promise in shielding bacteria until they reach intestinal destinations. Encapsulation within electrospun nanofibers protected probiotic strains far more effectively than standard buffers during storage and simulated digestion. Once within the gut, preprogrammed changes in acidity dissolve these nano-casings, precisely releasing probiotics. Such targeted delivery could greatly heighten viable probiotic concentrations contacting intestinal walls. Early human trials demonstrate 25 times more delivered bacteria surviving when encapsulated in pH-responsive polymer coatings. This could significantly impact probiotic activity in the body. Further possible innovation includes engineering programmable probiotic circuits and biosynthetic pathways. Using principles of synthetic biology, we may design 'smart' probiotics transcending natural limitations.

Probiotic circuits could sense specific gut metabolites, pH fluctuations, inflammation signals, or antibiotic concentrations. They might then launch tailored therapeutic responses like enhanced mucus production, toxin sequestering or anti-inflammatory signaling. Programming probiotics to produce custom metabolites is another frontier.

Bioengineered bacteria might yield targeted nutritional compounds, medications, specialized fatty acids or anti-inflammatory products. Integrating such micro-factories into the gut ecosystem could enable continuous in situ therapies. While safety concerns exist, active debate continues on responsible oversight of these transformative technologies. Beyond probiotics, metabolomic testing measuring microbial byproducts in serum could enable early, non-invasive disease detection. Dysfunctional gut communities often leak inflammatory metabolites underlying chronic inflammation. Metabolomics analyses changes in these circulating biomarkers, assessing systemic impacts. Already, research links inflammatory microbiome metabolites like TMAOq, indoxyl sulfate, and p-Cresol sulfate with heightened cardiovascular disease risk before symptoms manifest.

Ongoing studies investigate metabolomics profiling to predict Type 2 diabetes and mental health conditions. Catching chronic inflammation early with microbiome markers could improve prevention and outcomes enormously. Environmental metagenomics sampling shared surfaces may help reduce contagious disease transmission while supporting more targeted antibiotic use. Quickly distinguishing viral from bacterial infections helps avoid unnecessary antibiotics and antibiotic resistance. Recent innovations enabling rapid, real-time sequencing of microbial RNA from air and surfaces could aid infection control and stewardship efforts. Mapping resistance genes and pathogen strains in clinical settings has further potential to limit the spread of antibiotic-resistant microbes. Monitoring changes in surface biome populations could provide low-cost surveillance, enabling earlier interventions and limiting dangerous bacteria transmission between patients. Expanding environmental sampling in shared spaces offers much hope for improving infection prevention and control. Beyond probiotics, radically novel microbial manipulations show therapeutic promise.

For example, experimental phage therapy leverages viruses that attack specific bacteria to remove harmful strains selectively. Re-balancing runaway pathogens with targeted viruses offers tailored antimicrobial action without disturbing beneficial flora. While more research remains needed, early findings suggest phage therapy could succeed where antibiotics fail. Engineered phages defeated deadly CRE infections in animal models, and Compassionate Use cases report eliminating resistant bacterial infections in immunocompromised patients. As antibiotic resistance grows, virus-based microbial assassins may provide alternative solutions. Similar selectivity may come from emerging CRISPR gene editing therapies targeting deleterious disease-linked microbial strains. This stands for Clustered Regularly Interspaced Short Palindromic Repeats, the hallmark of a bacterial defense system that forms the basis for CRISPR-Cas9 genome editing technology. Early animal studies used CRISPR to delete virulence genes in pathogenic Candida albicans gut colonizers successfully.

Eliminating fungal genes enabling biofilm production and toxin secretion prevented fatal overgrowth. More research is needed, but this suggests microbiome gene editing could combat illness without removing beneficial microbes. CRISPR also shows promise in editing human genes and regulating microbiome interactions. Rare lysosomal storage disorders often manifest with severe gut inflammation and permeability from microbial dysbiosis. Mouse models suggest CRISPR correction of faulty human mucus or inflammation genes could reduce bacterial overgrowth and leakage driving systemic inflammation. Such complementary therapy could potentially improve outcomes for multiple chronic inflammatory conditions. While theoretical risks exist around the long-term ecological impacts of releasing gene-edited microbes, microbial gene editing remains an emerging discipline. Ongoing ethical discussion aims to ensure responsible oversight governing this microbiome frontier.

Perhaps the most radical area of microbiome medicine involves ecosystem-based approaches, seeing health as rooted in nurturing socio-ecological harmony at multiple interconnected levels. This model views balancing human interactions with microbial, social, and environmental ecosystems as key to wellness. For example, research reveals lifestyle, diet, and toxin exposures disrupt internal gut ecosystems, driving cellular dysfunction underlying mood disorders, autoimmunity, and metabolic disease. But these internal ecosystems intertwine deeply with external social, built, and natural environments.

Chronic stress, loneliness, indoor air quality, biodiversity contact, and environmental toxins converge, shaping internal ecological dynamics. Eco-system medicine thus seeks to align human activities and built environments, nurturing healthy internal/external ecology simultaneously. This includes dietary choices nurturing microbial diversity, affirming social connections, promoting regular outdoor biodiversity contact, removing home toxins, enabling stress resilience skills, and designing healthcare spaces to avoid ecological disruption. Aligning external ecosystems for optimal functioning of the internal microbiome, immune, and metabolic ecologies offers new dimensions for sustainable healing. An ecosystem lens sees societal choices like industrial food systems, sedentary built environments, inequities, and loneliness as disease vectors disrupting human ecology. This compels redesigning societies, cities, and healthcare investing in the commons underpinning public health. Cancer treatment represents another avenue for microbiome medicine. Growing evidence suggests gut microbes modulate responses to immune therapies like checkpoint inhibitors. Specific microbial metabolites like short-chain fatty acids, polysaccharide A, and secondary bile acids boost dendritic cell activity and anti-tumor lymphocytes. Germ-free mice show poorer tumor control under immunotherapy.

This links diverse, health-promoting gut microbes with stronger anti-cancer immune responses. Conversely, hungry, inflamed gut wall tissues stressed by dysbiosis spark immune suppression pathways counteracting checkpoint inhibitors. Oral microbiome disturbances and antibiotics are also associated with poorer survival outcomes after lung cancer immunotherapy. This suggests microbiome screening and correction before and during immuno-oncology offers much promise. Short courses of targeted antibiotics removing specific species that reduce efficacy, paired with high-fiber prebiotic diets-or possibly fecal transplants from high-responder donors-could potentially amplify survival. Ongoing research continues investigating best protocols. Additionally, emerging evidence hints that bacterial metabolite production affects therapy toxicity. Higher TMAO levels are associated with checkpoint inhibitor side effects, while macrophage activation by certain metabolites may protect the intestinal lining and blood-brain barrier integrity. This could influence adverse event screening and prevention. Further studies around microbes' impact on immuno-oncology safety are warranted.

Incredible potential also lies in leveraging gut microbes to positively influence mood and cognition as psychobiotic therapies. Multiple pathways exist whereby intestinal flora impact mental functioning. The vagus nerve provides direct gut-brain signaling, while microbial metabolites like short-chain fatty acids strengthen blood-brain integrity. Microbes also modulate neurotransmitters and their precursors. Certain probiotic strains demonstrate anxiety-reducing and stress-resilience impacts in early studies. For example, a formulation of L.Helveticus and Bifidobacterium longum taken for a month significantly improved depression, anxiety, and stress scales. L. rhamnosus also strengthened healthy emotional responses under stress by suppressing cortisol in healthy volunteers. Such psychobiotic supplements may benefit mood disorders, ADHD, autism, and schizophrenia.

However, microbiome connections likely also affect neurodegenerative conditions like Alzheimer's and Parkinson's. Inflammatory metabolic dysfunctions and impaired gut-brain signaling are associated with neurological decline. Protecting microbial communities early could support preventative neuro-protective strategies. Custom psychobiotic combinations may eventually allow personalized mental healthcare. Stool tests analyzing microbiome health could inform probiotic pairings tailored for unique neurotransmitter needs based on gene panels. Microbe-driven neurochemistry varies greatly between individuals. Designing microbiome therapies aligning with personal nervous system biology offers much promise for mind-body healing.

While still early, gut ecology may rank among the most consequential medical revelations in human history. Microbes' profound role in human health and disease irrevocably reshapes medicine's conceptual foundations. And our species' fate increasingly depends upon how we engage this hidden universe within. Further illuminating mechanisms driving microbiome/health interdependencies can unlock tremendous healing potential. But realizing possibilities requires expanding science beyond siloed reductionism and appreciating contextual eco-social complexities. Vibrant futures lie in sustaining the deep microbial connections linking human and environmental flourishing. Should we nurture the fertile symbioses sustaining all life, a new era of possibilities shimmers ahead. Humanity has reached a critical juncture with both microbiome crisis and microbiome revival. Industrial diets, chemicals, and schools separated from nature disrupt children's microbiome development, feeding 21st-century disease epidemics.

Yet healthier paths exist. Choices enabling naturalistic microbiome cultivation early and often can profoundly shape life-long resilience and balance inflammation underlying chronic illness. The rest is integration. Rest remembers our kinship with the deeply shared microbial ecosystems coding life's poetry across the Earth. Human health is community health. And here, now emerges an opportunity for microbial cooperation, for symbiotic repair of bodies and biospheres. May this book kindle remembrance of the key roles microbes play in sustaining wellness. And from this vision, let cultures flourish, celebrating how nurturing tiny partners nourishes us, societies, and the living planet. Venturing into microbiome medicine's future leads us imaginatively into a world of diverse microbial therapies and diagnostics.

Patients could soon swallow 'smart' personalized genetically engineered probiotics programmed to sense and respond to internal imbalances. Microbiome tests diagnosing subtypes and vitals may guide dietary and lifestyle changes required to restore balance within each person's unique inner ecology. Meanwhile, hospitals could employ advanced instrumentation surveying environmental microbiomes, predicting where infections flourish before thriving microbes colonize immunosuppressed patients. Throughout this book, we have learned extensively about the fascinating microbiomes keeping us alive and well. We have explored the intricate systems of bacteria and fungi that have developed over eons and have evolved to deal with a rapidly changing earth and environment. The impact of the Covid-19 pandemic has made many more people interested in bacteria and the way it works both inside and outside our bodies. The complexity of it all is enthralling – and globally, skilled researchers and scientists are trying to put the puzzle pieces of this wondrous jigsaw together. As we learn and develop, our challenges are inevitably immense. Together with empathy and understanding, we can all greatly respect the health challenges facing our communities!

www.ingramcontent.com/pod-product-compliance
Lightning Source LLC
Chambersburg PA
CBHW071349230926
51846CB00002B/483
* 9 7 8 9 4 9 3 3 7 8 4 6 9 *